Electricity And Matter

Joseph J. Thomson

ELECTRICITY AND MATTER

BY

J. J. THOMSON, D.Sc., LL.D., Ph.D., F.R.S.

FELLOW OF TRINITY COLLEGE, CAMBRIDGE; CAVENDISH
PROFESSOR OF EXPERIMENTAL PHYSICS, CAMBRIDGE

WITH DIAGRAMS

WESTMINSTER
ARCHIBALD CONSTABLE & CO., Ltd.
1904

Printed by the Trow Directory, Printing and Bookbinding Company
New York, U. S. A.

THE SILLIMAN FOUNDATION.

In the year 1883 a legacy of eighty thousand dollars was left to the President and Fellows of Yale College in the city of New Haven, to be held in trust, as a gift from her children, in memory of their beloved and honored mother Mrs. Hepsa Ely Silliman.

On this foundation Yale College was requested and directed to establish an annual course of lectures designed to illustrate the presence and providence, the wisdom and goodness of God, as manifested in the natural and moral world. These were to be designated as the Mrs. Hepsa Ely Silliman Memorial Lectures. It was the belief of the testator that any orderly presentation of the facts of nature or history contributed to the end of this foundation more effectively than any attempt to emphasize the elements of doctrine or of creed; and he therefore provided that lectures on dogmatic or polemical theology should be excluded from the scope of this foundation, and that the subjects should be selected rather from the domains of natural science and history, giving special prominence to astronomy, chemistry, geology, and anatomy.

It was further directed that each annual course should be made the basis of a volume to form part of a series constituting a memorial to Mrs. Silliman. The memorial fund came into the possession of the Corporation of Yale University in the year 1902; and the present volume constitutes the first of the series of memorial lectures.

PREFACE

In these Lectures given at Yale University in May, 1903, I have attempted to discuss the bearing of the recent advances made in Electrical Science on our views of the Constitution of Matter and the Nature of Electricity; two questions which are probably so intimately connected, that the solution of the one would supply that of the other. A characteristic feature of recent Electrical Researches, such as the study and discovery of Cathode and Röntgen Rays and Radio-active Substances, has been the very especial degree in which they have involved the relation between Matter and Electricity.

In choosing a subject for the Silliman Lectures, it seemed to me that a consideration of the bearing of recent work on this relationship might be suitable, especially as such a discussion suggests multitudes of questions which would furnish admirable subjects for further investigation by some of my hearers.

Cambridge, Aug., 1903.

J. J. THOMSON.

CONTENTS

ELECTRICITY AND MATTER

CHAPTER I

REPRESENTATION OF THE ELECTRIC FIELD BY LINES OF FORCE

My object in these lectures is to put before you in as simple and untechnical a manner as I can some views as to the nature of electricity, of the processes going on in the electric field, and of the connection between electrical and ordinary matter which have been suggested by the results of recent investigations.

The progress of electrical science has been greatly promoted by speculations as to the nature of electricity. Indeed, it is hardly possible to overestimate the services rendered by two theories as old almost as the science itself; I mean the theories known as the two- and the one-fluid theories of electricity.

The two-fluid theory explains the phenomena of electro-statics by supposing that in the universe there are two fluids, uncreatable and indestructible, whose presence gives rise to electrical effect one of these fluids is called positive, the oth negative electricity, and electrical phenome are explained by ascribing to the fluids the fo lowing properties. The particles of the positi fluid repel each other with forces varying inverse as the square of the distance between them, as d also the particles of the negative fluid; on th other hand, the particles of the positive fluid a tract those of the negative fluid. The attracti between two charges, m and m', of opposite sig are in one form of the theory supposed to b exactly equal to the repulsion between tw charges, m and m' of the same sign, placed i the same position as the previous charges. In other development of the theory the attraction supposed to slightly exceed the repulsion, so as afford a basis for the explanation of gravitation

The fluids are supposed to be exceedingly m bile and able to pass with great ease through c ductors. The state of electrification of a bod determined by the *difference* between the qu ties of the two electric fluids contained by it; contains more positive fluid than negative i positively electrified, if it contains equal quanti it is uncharged. Since the fluids are uncreat

— our home of 47 y

all that goes with it

) has not been easy.

es that are presently

20, has nicely settle

B.C. V6J 4H2.

good form. Gigi an

oth continue to encha

cept, will be publishe

and indestructible, the appearance of the positive fluid in one place must be accompanied by the departure of the same quantity from some other place, so that the production of electrification of one sign must always be accompanied by the production of an equal amount of electrification of the opposite sign.

On this view, every body is supposed to consist of three things: ordinary matter, positive electricity, negative electricity. The two latter are supposed to exert forces on themselves and on each other, but in the earlier form of the theory no action was contemplated between ordinary matter and the electric fluids; it was not until a comparatively recent date that Helmholtz introduced the idea of a specific attraction between ordinary matter and the electric fluids. He did this to explain what is known as contact electricity, *i.e.*, the electrical separation produced when two metals, say zinc and copper, are put in contact with each other, the zinc becoming positively, the copper negatively electrified. Helmholtz supposed that there are forces between ordinary matter and the electric fluids varying for different kinds of matter, the attraction of zinc for positive electricity being greater than that of copper, so that when these metals are put in contact the zinc robs the copper of some of its positive electricity.

There is an indefiniteness about the two-fluid theory which may be illustrated by the consideration of an unelectrified body. All that the two-fluid theory tells us about such a body is that it contains equal quantities of the two fluids. It gives no information about the amount of either; indeed, it implies that if equal quantities of the two are added to the body, the body will be unaltered, equal quantities of the two fluids exactly neutralizing each other. If we regard these fluids as being anything more substantial than the mathematical symbols + and − this leads us into difficulties; if we regard them as physical fluids, for example, we have to suppose that the mixture of the two fluids in equal proportions is something so devoid of physical properties that its existence has never been detected.

The other fluid theory—the one-fluid theory of Benjamin Franklin—is not open to this objection. On this view there is only one electric fluid, the positive; the part of the other is taken by ordinary matter, the particles of which are supposed to repel each other and attract the positive fluid, just as the particles of the negative fluid do on the two-fluid theory. Matter when unelectrified is supposed to be associated with just so much of the electric fluid that the attraction of the matter on a portion of the electric fluid outside it is just sufficient to counteract the repulsion exerted on the same fluid by the electric fluid associated with the matter. On this view, if the quantity of matter in a body is known the quantity of electric fluid is at once determined.

The services which the fluid theories have rendered to electricity are independent of the notion of a fluid with any physical properties; the fluids were mathematical fictions, intended merely to give a local habitation to the attractions and repulsions existing between electrified bodies, and served as the means by which the splendid mathematical development of the theory of forces varying inversely as the square of the distance which was inspired by the discovery of gravitation could be brought to bear on electrical phenomena. As long as we confine ourself to questions which only involve the law of forces between electrified bodies, and the simultaneous production of equal quantities of + and − electricity, both theories must give the same results and there can be nothing to decide between them. The physicists and mathematicians who did most to develop the "Fluid Theories" confined themselves to questions of this kind, and refined and idealized the conception of these fluids until any reference to their physical properties was considered almost indelicate. It is not until we investigate phenomena which involve the physical properties of the fluid that we can hope to distinguish between the rival fluid theories. Let us take a case which has actually arisen. We have been able to measure the masses associated with given charges of electricity in gases at low pressures, and it has been found that the mass associated with a positive charge is immensely greater than that associated with a negative one. This difference is what we should expect on Franklin's one-fluid theory, if that theory were modified by making the electric fluid correspond to negative instead of positive electricity, while we have no reason to anticipate so great a difference on the two-fluid theory. We shall, I am sure, be struck by the similarity between some of the views which we are led to take by the results of the most recent researches with those enunciated by Franklin in the very infancy of the subject.

Faraday's Line of Force Theory

The fluid theories, from their very nature, imply the idea of action at a distance. This idea, although its convenience for mathematical analysis has made it acceptable to many mathematicians, is one which many of the greatest physicists have felt utterly unable to accept, and have devoted much thought and labor to replacing it by something involving mechanical continuity. Pre-eminent among them is Faraday. Faraday was deeply influenced by the axiom, or if you prefer it, dogma that matter cannot act where it is not. Faraday, who possessed, I believe, almost unrivalled mathematical insight, had had no training in analysis, so that the convenience of the idea of action at a distance for purposes of calculation had no chance of mitigating the repugnance he felt to the idea of forces acting far away from their base and with no physical connection with their origin. He therefore cast about for some way of picturing to himself the actions in the electric field which would get rid of the idea of action at a distance, and replace it by one which would bring into prominence some continuous connection between the bodies exerting the forces. He was able to do this by the conception of lines of force. As I shall have continually to make use of this method, and as I believe its powers and possibilities have never been adequately realized, I shall devote some time to the discussion and development of this conception of the electric field.

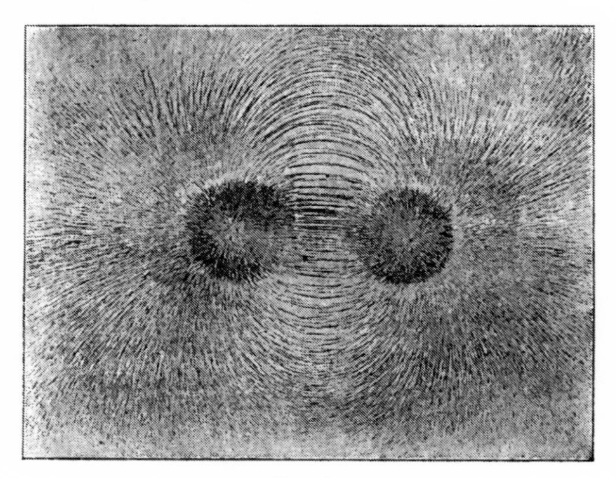

Fig. 1.

The method was suggested to Faraday by the consideration of the lines of force round a bar magnet. If iron filings are scattered on a smooth surface near a magnet they arrange themselves as in Fig. 1; well-marked lines can be traced running from one pole of the magnet to the other; the direction of these lines at any point coincides with the direction of the magnetic force, while the intensity of the force is indicated by the concentration of the lines. Starting from any point in the field and travelling always in the direction of the magnetic force, we shall trace out a line which will not stop until we reach the negative pole of the magnet; if such lines are drawn at all points in the field, the space through which the magnetic field extends will be filled with a system of lines, giving the space a fibrous structure like that possessed by a stack of hay or straw, the grain of the structure being along the lines of force. I have spoken so far only of lines of magnetic force; the same considerations will apply to the electric field, and we may regard the electric field as full of lines of electric force, which start from positively and end on negatively electrified bodies. Up to this point the process has been entirely geometrical, and could have been employed by those who looked at the question from the point of view of action at a distance; to Faraday, however, the lines of force were far more than mathematical abstractions—they were physical realities. Faraday materialized the lines of force and endowed them with physical properties so as to explain the phenomena of the electric field. Thus he supposed that they were in a state of tension, and that they repelled each other. Instead of an intangible action at a distance between two electrified bodies, Faraday regarded the whole space between the bodies as full of stretched mutually repellent springs. The charges of electricity to which alone an interpretation had been given on the fluid theories of electricity were on this view just the ends of these springs, and an electric charge, instead of being a portion of fluid confined to the electrified body, was an extensive arsenal of springs spreading out in all directions to all parts of the field.

To make our ideas clear on this point let us consider some simple cases from Faraday's point of view. Let us first take the case of two bodies with equal and opposite charges, whose lines of force are shown in Fig. 2. You notice that the lines of force are most dense along $A B$, the line joining the bodies, and that there are more lines of force on the side of A nearest to B than on the opposite side. Consider the effect of the lines of force on A; the lines are in a state of tension and are pulling away at A; as there are more pulling at A on the side nearest to B than on the opposite side, the pulls on A toward B overpower those pulling A away from B, so that A will tend to move toward B; it was in this way that Faraday pictured to himself the attraction between oppositely electrified bodies. Let us now consider the condition of one of the curved lines of force, such as PQ; it is in a state

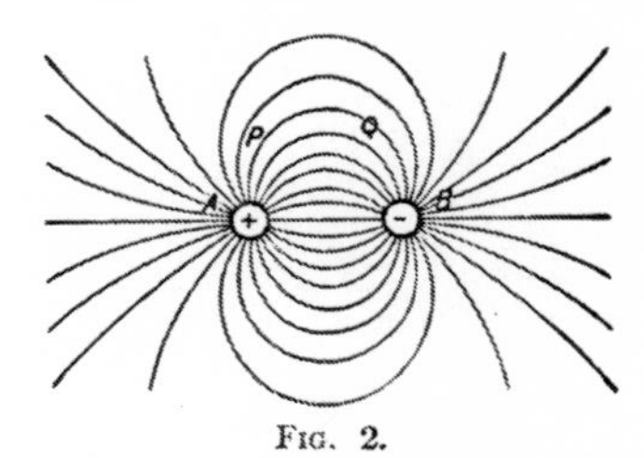

Fig. 2.

of tension and will therefore tend to straighten itself, how is it prevented from doing this and maintained in equilibrium in a curved position? We can see the reason for this if we remember that the lines of force repel each other and that the lines are more concentrated in the region between PQ and AB than on the other side of PQ; thus the repulsion of the lines inside PQ will be greater than the repulsion of those outside and the line PQ will be bent outwards.

Let us now pass from the case of two oppositely electrified bodies to that of two similarly electrified ones, the lines of force for which are shown in Fig. 3. Let us suppose A and B are positively electrified; since the lines of force start from positively and end on negatively electrified bodies, the lines starting from A and B will travel away to join some body or bodies possessing the

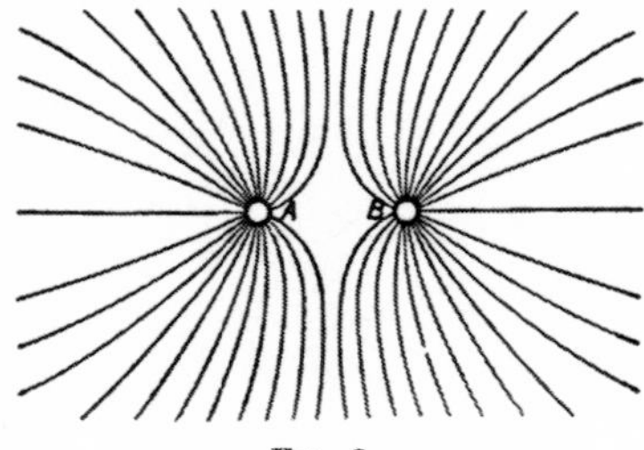

Fig. 3.

negative charges corresponding to the positive ones on A and B; let us suppose that these charges are a considerable distance away, so that the lines of force from A would, if B were not present, spread out, in the part of the field under consideration, uniformly in all directions. Consider now the effect of making the system of lines of force attached to A and B approach each other;

since these lines repel each other the lines of force on the side of A nearest B will be pushed to the opposite side of A, so that the lines of force will now be densest on the far side of A; thus the pulls exerted on A in the rear by the lines of force will be greater than those in the front and the result will be that A will be pulled away from B. We notice that the mechanism producing this repulsion is of exactly the same type as that which produced the attraction in the previous case, and we may if we please regard the repulsion between A and B as due to the attractions on A and B of the complementary negative charges which must exist in other parts of the field.

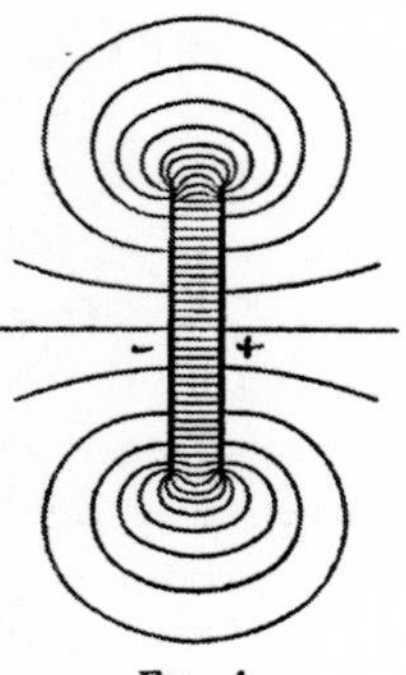

Fig. 4.

The results of the repulsion of the lines of force are clearly shown in the case represented in Fig. 4, that of two oppositely electrified plates; you will notice that the lines of force between the plates are straight except near the edges of the plates; this is what we should expect as the downward pressure exerted by the lines of force above a line in this part of the field will be equal to the upward pressure exerted by those below it. For a line of force near the edge of the plate, however, the pressure of the lines of force below will exceed the pressure from those above, and the line of force will bulge out until its curvature and tension counteract the squeeze from inside; this bulging is very plainly shown in Fig. 4.

So far our use of the lines of force has been descriptive rather than metrical; it is, however, easy to develop the method so as to make it metrical. We can do this by introducing the idea of *tubes of force.* If through the boundary of any small closed curve in the electric field we draw the lines of force, these lines will form a tubular surface, and if we follow the lines back to the positively electrified surface from which they start and forward on to the negatively electrified surface on which they end, we can prove that the positive charge enclosed by the tube at its origin is equal to the negative charge enclosed by it at its end. By properly choosing the area of the small curve through which we draw the lines of force, we may arrange that the charge enclosed by the tube is equal to the unit charge. Let us call such a tube a Faraday tube—then each unit of positive electricity in the field may be regarded as the origin and each unit of negative electricity as the termination of a Faraday tube. We regard these Faraday tubes as having direction, their direction being the same as that of the electric force, so that the positive direction is from the positive to the negative end of the tube. If we draw any closed surface then the difference between the number of Faraday tubes which pass out of the surface and those which pass in will be equal to the algebraic sum of the charges inside the surface; this sum is what Maxwell called the *electric displacement* through the surface. What Maxwell called the *electric displacement in any direction at a point* is the number of Faraday tubes which pass through a unit area through the point drawn at right angles to that direction, the number being reckoned algebraically; *i.e.*, the lines which pass through in one direction being taken as positive, while those which pass through in the opposite direction are taken as negative, and the number passing through the area is the difference between the number passing through positively and the number passing through negatively.

For my own part, I have found the conception

of Faraday tubes to lend itself much more readily to the formation of a mental picture of the processes going on in the electric field than that of electric displacement, and have for many years abandoned the latter method.

Maxwell took up the question of the tensions and pressures in the lines of force in the electric field, and carried the problem one step further than Faraday. By calculating the amount of these tensions he showed that the mechanical effects in the electrostatic field could be explained by supposing that each Faraday tube force exerted a tension equal to R, R being the intensity of the electric force, and that, in addition to this tension, there was in the medium through which the tubes pass a hydrostatic pressure equal to $\frac{1}{2}NR$, N being the density of the Faraday tubes; *i.e.*, the number passing through a unit area drawn at right angles to the electric force. If we consider the effect of these tensions and pressure on a unit volume of the medium in the electric field, we see that they are equivalent to a tension $\frac{1}{2}NR$ along the direction of the electric force and an equal pressure in all directions at right angles to that force.

Moving Faraday Tubes

Hitherto we have supposed the Faraday tubes to be at rest, let us now proceed to the study of the effects produced by the motion of those tubes. Let us begin with the consideration of a very simple case—that of two parallel plates, A and B, charged, one with positive the other with negative electricity, and suppose that after being charged the plates are connected by a conducting wire, EFG. This wire will pass through some of the outlying tubes; these tubes, when in a conductor, contract to molecular dimensions and the repulsion they previously exerted on neighboring tubes will therefore disappear. Consider the effect of this on a tube PQ between the plates; PQ was originally in equilibrium under its own tension, and the repulsion exerted by the neighboring tubes. The repulsions due to those cut by EFG have now, however, disappeared so that PQ will no longer be in equilibrium, but will be pushed towards EFG. Thus, more and more tubes will be pushed into EFG, and we shall

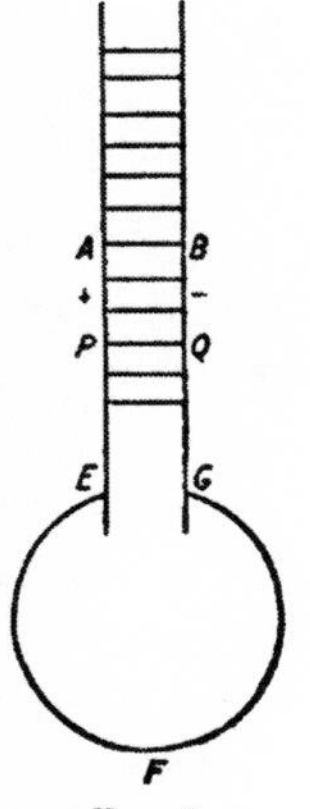

FIG. 5.

have a movement of the whole set of tubes between the plates toward EFG. Thus, while the discharge of the plates is going on, the tubes between the plates are moving at right angles to themselves. What physical effect accompanies this movement of the tubes? The result of connecting the plates by EFG is to produce a current of electricity flowing from the positively charged plate through EFG to the negatively charged plate; this is, as we know, accompanied by a magnetic force between the plates. This magnetic force is at right angles to the plane of the paper and equal to 4π times the intensity of the current in the plate, or, if σ is the density of the charge of electricity on the plates and v the velocity with which the charge moves, the magnetic force is equal to $4\pi\sigma v$.

Here we have two phenomena which do not take place in the steady electrostatic field, one the movement of the Faraday tubes, the other the existence of a magnetic force; this suggests that there is a connection between the two, and that motion of the Faraday tubes is accompanied by the production of magnetic force. I have followed up the consequences of this supposition and have shown that, if the connection between the magnetic force and the moving tubes is that given below, this view will account for Ampère's laws connecting current and magnetic force, and for Faraday's law of the induction of currents. Maxwell's great contribution to electrical theory, that variation in the electric displacement in a dielectric produces magnetic force, follows at once from this view. For, since the electric displacement is measured by the density of the Faraday tubes, if the electric displacement at any place changes, Faraday tubes must move up to or away from the place, and motion of Faraday tubes, by hypothesis, implies magnetic force.

The law connecting magnetic force with the motion of the Faraday tubes is as follows: A Faraday tube moving with velocity v at a point P, produces at P a magnetic force whose magnitude is $4\pi v \sin \theta$, the direction of the magnetic force being at right angles to the Faraday tube, and also to its direction of motion; θ is the angle between the Faraday tube and the direction in which it is moving. We see that it is only the motion of a tube at right angles to itself which produces magnetic force; no such force is produced by the gliding of a tube along its length.

Motion of a Charged Sphere

We shall apply these results to a very simple and important case—the steady motion of a charged sphere. If the velocity of the sphere is small compared with that of light then the Faraday tubes will, as when the sphere is at rest, be uniformly distributed and radial in direction. They will be carried along with the sphere. If e is the charge on the sphere, O its centre, the density of the Faraday tubes at P is $\frac{e}{4\pi\, OP^2}$; so that if v is the velocity of the sphere, θ the

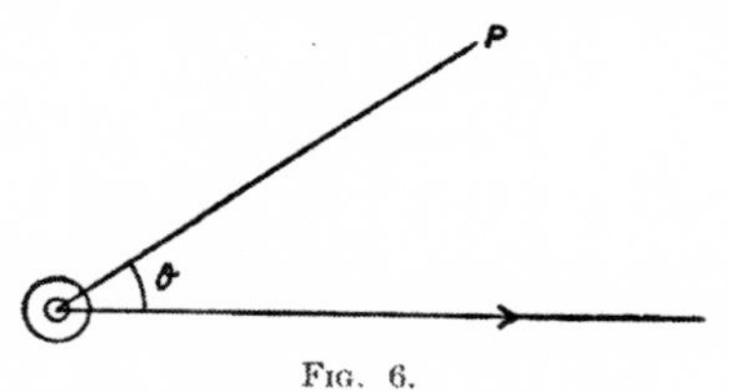

Fig. 6.

angle between OP and the direction of motion of the sphere, then, according to the above rule, the magnetic force at P will be $\frac{e v \sin\theta}{r^2}$, the direction of the force will be at right angles to OP, and at right angles to the direction of motion of the sphere; the lines of magnetic force will thus be circles, having their centres on the path of the centre of the sphere and their planes at right angles to this path. Thus, a moving charge of electricity will be accompanied by a magnetic field. The existence of a magnetic field implies energy; we know that in a unit volume of the field at a place where the magnetic force is H there are $\frac{\mu H^2}{8\pi}$ units of energy, where μ is the magnetic permeability of the medium. In the case of the moving sphere the energy per unit volume at P is $\frac{\mu e^2 v^2 \sin^2 \theta}{8\pi \, OP^4}$. Taking the sum of this energy for all parts of the field outside the sphere, we find that it amounts to $\frac{\mu e^2 v^2}{3a}$, where a is the radius of the sphere. If m is the mass of the sphere, the kinetic energy in the sphere is $\frac{1}{2} m v^2$; in addition to that we have the energy outside the sphere, which as we have seen is $\frac{\mu e^2 v^2}{3a}$; so that the whole kinetic energy of the system is $\frac{1}{2}\left(m + \frac{2\mu}{3}\frac{e^2}{a}\right) v^2$, or the energy is the same as if the mass of the sphere were $m + \frac{2\mu}{3}\frac{e^2}{a}$ instead of m. Thus, in consequence of the electric charge, the mass of the sphere is measured by $\frac{2\mu e^2}{3a}$. This is a very important result, since it shows that part of the mass of a charged sphere is due to its charge. I shall later on have to bring before you considerations which show that it is not impossible that the whole mass of a body may arise in the way.

Before passing on to this point, however, I should like to illustrate the increase which takes place in the mass of the sphere by some analogies drawn from other branches of physics. The first of these is the case of a sphere moving through a frictionless liquid. When the sphere moves it sets the fluid around it moving with a velocity proportioned to its own, so that to move the sphere we we have not merely to move the substance of the sphere itself, but also the liquid around it; the consequence of this is, that the sphere behaves as if its mass were increased by that of a certain volume of the liquid. This volume, as was shown by Green in 1833, is half the volume of the sphere. In the case of a cylinder moving at right angles to its length, its mass is increased by the mass of an equal volume of the liquid. In the case of an elongated body like a cylinder, the amount by which the mass is increased depends upon the direction in which the body is moving, being much smaller when the body moves point foremost than when moving sideways. The mass of such a body depends on the direction in which it is moving.

Let us, however, return to the moving electrified sphere. We have seen that in consequence of its charge its mass is increased by $\frac{2\mu e^2}{3a}$; thus, if it is moving with the velocity v, the momentum is not mv, but $\left(m + \frac{2\mu e^2}{3a}\right)v$. The additional momentum $\frac{2\mu e^2}{3a} v$ is not in the sphere, but in the space surrounding the sphere. There is in this space *ordinary mechanical momentum*, whose resultant is $\frac{2\mu e^2}{3a} v$ and whose direction is parallel to the direction of motion of the sphere. It is important to bear in mind that this momentum is not in any way different from ordinary mechanical momentum and can be given up to or taken from the momentum of moving bodies. I want to bring the existence of this momentum before you as vividly and forcibly as I can, because the recognition of it makes the behavior of the electric field entirely analogous to that of a mechanical system. To take an example, according to Newton's Third Law of Motion, Action and Reaction are equal and opposite, so that the momentum in any direction of any self-contained system is invariable. Now, in the case of many electrical systems there are apparant violations of this principle; thus, take the case of a charged body at rest struck by an electric pulse, the charged body when exposed to the electric force in the pulse acquires velocity and momentum, so that when the pulse has passed over it, its momentum is not what it was originally. Thus, if we confine our attention to the momentum in the charged body, *i.e.*, if we suppose that momentum is necessarily confined to what we consider ordinary matter, there has been a violation of the Third Law of Motion, for the only momentum recognized on this restricted view has been changed. The phenomenon is, however, brought into accordance with this law if we recognize the existence of the momentum in the electric field; for, on this view, before the pulse reached the charged body there was momentum in the pulse, but none in the body; after the pulse passed over the body there was some momentum in the body and a smaller amount in the pulse, the loss of momentum in the pulse being equal to the gain of momentum by the body.

We now proceed to consider this momentum more in detail. I have in my "Recent Researches on Electricity and Magnetism" calculated the amount of momentum at any point in the electric field, and have shown that if N is the number of Faraday tubes passing through a unit area drawn at right angles to their direction, B the magnetic induction, θ the angle between the induction and the Faraday tubes, then the momentum per unit volume is equal to $N B \sin \theta$, the direction of the momentum being at right angles to the magnetic induction and also to the Faraday tubes. Many of you will notice that the momentum is parallel to what is known as Poynting's vector—the vector whose direction gives the direction in which energy is flowing through the field.

Moment of Momentum Due to an Electrified Point and a Magnetic Pole

To familiarize ourselves with this distribution of momentum let us consider some simple cases in detail. Let us begin with the simplest, that of an electrified point and a magnetic pole; let A, Fig. 7, be the point, B the pole. Then, since the momentum at any point P is at right angles to $A\ P$, the direction of the Faraday tubes and also to $B\ P$, the magnetic induction, we see that the momentum will be perpendicular to the plane $A\ B\ P$; thus, if we draw a series of lines such that their direction at any point coincides with the direction of the momentum at that point, these lines will form a series of circles whose planes are perpendicular to the line $A\ B$, and whose centres lie along that line. This distribution of momentum, as far as direction goes, is that possessed by a top spinning around $A\ B$. Let us now find what this distribution of momentum throughout the field is equivalent to.

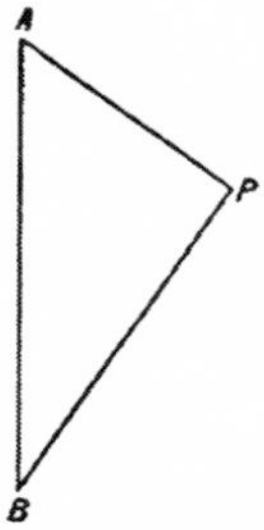

FIG. 7.

It is evident that the resultant momentum in any direction is zero, but since the system is spinning round $A\ B$, the direction of rotation being everywhere the same, there will be a finite moment of momentum round $A\ B$. Calculating the value of this from the expression for the momentum given above, we obtain the very simple expression $e\,m$ as the value of the moment of momentum about $A\ B$, e being the charge on the point and m the strength of the pole. By means of this expression we can at once find the moment of momentum of any distribution of electrified points and magnetic poles.

To return to the system of the point and pole, this conception of the momentum of the system leads directly to the evaluation of the force acting on a moving electric charge or a moving magnetic pole. For suppose that in the time δt the electrified point were to move from A to A', the moment of momentum is still $e m$, but its axis is along $A' B$ instead of $A B$. The moment of momentum of the field has thus changed, but the whole moment of momentum of the system comprising point, pole, and field must be constant, so that the change in the moment of momentum of the field must be accompanied by an equal and opposite change in the moment of momentum of the pole and point. The momentum gained by the point must be equal and opposite to that gained by the pole, since the whole momentum is zero. If θ is the angle $A B A'$, the change in the moment of momentum is $e m \sin \theta$, with an axis at right angles to $A B$ in the plane of the paper. Let δI be the change in the momentum of A, —

FIG. 8.

δI that of B, then δI and $-\delta I$ must be equivalent to a couple whose axis is at right angles to $A B$ in the plane of the paper, and whose moment is $e m \sin \theta$. Thus δI must be at right angles to the plane of the paper and

$$\delta I . A B = e m \sin \theta = \frac{e m A A' \sin \phi}{A B}$$

Where ϕ is the angle $B A A'$. If v is the velocity of A, $A A' = v \delta t$ and we get

$$\delta I = \frac{e m v \sin \phi \delta t}{A B^2}$$

This change in the momentum may be supposed due to the action of a force F perpendicular to the plane of the paper, F being the rate of increase of the momentum, or $\frac{\delta I}{\delta t}$. We thus get $F = \frac{e m v \sin \phi}{A B^2}$; or the point is acted on by a force equal to e multiplied by the component of the magnetic force at right angles to the direction of motion. The direction of the force acting on the point is at right angles to its velocity and also to the magnetic force. There is an equal and opposite force acting on the magnetic pole.

The value we have found for F is the ordinary expression for the mechanical force acting on a moving charged particle in a magnetic field; it may be written as $e v H \sin \phi$, where H's is the strength of the magnetic field. The force acting on unit charge is therefore $v H \sin \phi$. This mechanical force may be thus regarded as arising from an electric force $v H \sin \phi$, and we may express the result by saying that when a charged body is moving in a magnetic field an electric force $v H \sin \phi$ is produced. This force is the well-known electromotive force of induction due to motion in a magnetic field.

The forces called into play are due to the *relative* motion of the pole and point; if these are moving with the same velocity, the line joining them will not alter in direction, the moment of momentum of the system will remain unchanged and there will not be any forces acting either on the pole or the point.

The distribution of momentum in the system of pole and point is similar in some respects to that in a top spinning about the line $A B$. We can illustrate the forces acting on a moving electrified body by the behavior of such a top. Thus, let Fig. 9 represent a balanced gyroscope spinning about the axis $A B$, let the ball at A represent the electrified point, that at B the magnetic pole. Suppose the instrument is spinning with $A B$

horizontal, then if with a vertical rod I push against $A\ B$ horizontally, the point A will not merely move horizontally forward in the direction in which it is pushed, but will also move vertically upward or downward, just as a charged

FIG. 9.

point would do if pushed forward in the same way, and if it were acted upon by a magnetic pole at B.

Maxwell's Vector Potential

There is a very close connection between the momentum arising from an electrified point and a magnetic system, and the Vector Potential of that system, a quantity which plays a very large part in Maxwell's Theory of Electricity. From the expression we have given for the moment of momentum due to a charged point and a magnetic pole, we can at once find that due to a charge e of electricity at a point P, and a little magnet $A\,B$; let the negative pole of this magnet be at A, the positive at B, and let m be the strength of either pole. A simple calculation shows that in this case the axis of the resultant moment of momentum is in the plane $P\,A\,B$ at right angles to $P\,O$, O being the middle point of $A\,B$, and that the magnitude of the moment of momentum is equal to $e.\,m.\,A\,B\,\frac{\sin\phi}{OP^2}$, where ϕ is the angle $A\,B$ makes with $O\,P$. This moment of momentum is equivalent in direction and magnitude to that due to a momentum $e.\,m.\,A\,B\,\frac{\sin\phi}{OP^2}$ at P directed at right angles to the plane $P\,A\,B$, and another momentum equal in magnitude and opposite in direction at O. The vector $m\,A\,B\,\frac{\sin\phi}{OP^2}$ at P at right angles to the plane $P\,A\,B$ is the vector called by Maxwell the Vector Potential at P due to the Magnet.

Calling this Vector Potential I, we see that the momentum due to the charge and the magnet is equivalent to a momentum eI at P and a momentum $-eI$ at the magnet.

We may evidently extend this to any complex system of magnets, so that if I is the Vector Potential at P of this system, the momentum in the field is equivalent to a momentum eI at P together with momenta at each of the magnets equal to

$-e$ (Vector Potential at P due to that magnet).

If the magnetic field arises entirely from electric currents instead of from permanent magnets, the momentum of a system consisting of an electrified point and the currents will differ in some of its features from the momentum when the magnetic field is due to permanent magnets. In the latter case, as we have seen, there is a moment of momentum, but no resultant momentum. When, however, the magnetic field is entirely due to electric currents, it is easy to show that there is a resultant momentum, but that the moment of momentum about any line passing through the electrified particle vanishes. A simple calculation shows that the whole momentum in the field is equivalent to a momentum eI at the electrified point I being the Vector Potential at P due to the currents.

Thus, whether the magnetic field is due to permanent magnets or to electric currents or partly to one and partly to the other, the momentum when an electrified point is placed in the field at P is equivalent to a momentum $e\,I$ at P where I is the Vector Potential at P. If the magnetic field is entirely due to currents this is a complete representation of the momentum in the field; if the magnetic field is partly due to magnets we have in addition to this momentum at P other momenta at these magnets; the magnitude of the momentum at any particular magnet is $-e$ times the Vector Potential at P due to that magnet.

The well-known expressions for the electromotive forces due to Electro-magnetic Induction follow at once from this result. For, from the Third Law of Motion, the momentum of any self-contained system must be constant. Now the momentum consists of (1) the momentum in the field; (2) the momentum of the electrified point, and (3) the momenta of the magnets or circuits carrying the currents. Since (1) is equivalent to a momentum $e\,I$ at the electrified particle, we see that changes in the momentum of the field must be accompanied by changes in the momentum of the particle. Let M be the mass of the electrified particle, u, v, w the components parallel to the axes of x, y, z of its velocity, F, G, H, the components parallel to these axes of the Vector Potential at P, then the momentum of the field is equivalent to momenta eF, eG, eH at P parallel to the axes of x, y, z; and the momentum of the charged point at P has for components Mu, Mo, Mw. As the momentum remains constant, $Mu + eF$ is constant, hence if δu and δF are simultaneous changes in u and F,

$$M\delta u + e\delta F = 0;$$

$$\text{or } m\frac{du}{dt} = -e\frac{dF}{dt}.$$

From this equation we see that the point with the charge behaves as if it were acted upon by a mechanical force parallel to the axis of x and equal to $-e\frac{dF}{dt}$, *i.e.*, by an electric force equal to $-\frac{dF}{dt}$. In a similar way we see that there are electric forces $-\frac{dG}{dt}$, $-\frac{dH}{dt}$, parallel to y and z respectively. These are the well-known expressions of the forces due to electro-magnetic induction, and we see that they are a direct consequence of the principle that action and reaction are equal and opposite.

Readers of Faraday's Experimental Researches will remember that he is constantly referring to what he called the "Electrotonic State"; thus he regarded a wire traversed by an electric current as being in the Electrotonic State when in a magnetic field. No effects due to this state can be detected as long as the field remains constant; it is when it is changing that it is operative. This Electrotonic State of Faraday is just the *momentum* existing in the field.

CHAPTER II

ELECTRICAL AND BOUND MASS.

I WISH in this chapter to consider the connection between the momentum in the electric field and the Faraday tubes, by which, as I showed in the last lecture, we can picture to ourselves the state of such a field. Let us begin by considering the case of the moving charged sphere. The lines of electric force are radial; those of magnetic force are circles having for a common axis the line of motion of the centre of the sphere; the momentum at a point P is at right angles to each of these directions and so is at right angles to $O P$ in the plane containing P and the line of motion of the centre of the sphere.

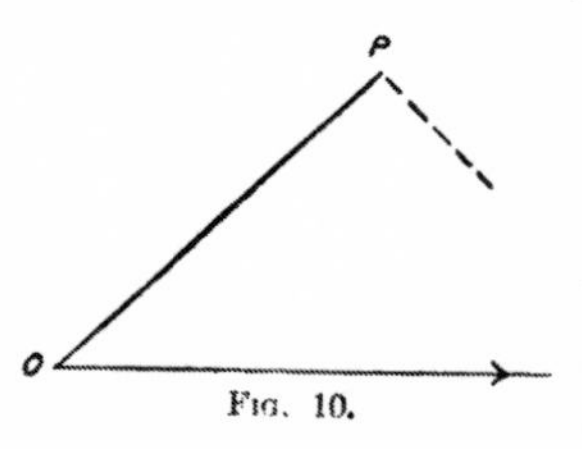

FIG. 10.

If the number of Faraday tubes passing through a unit area at P placed at right angles to $O P$ is N, the magnetic induction at P is, if μ is the magnetic permeability of the medium surrounding the sphere, $4\pi\mu Nv \sin\theta$, v being the velocity of the sphere and θ the angle OP makes with the direction of motion of the sphere. By the rule given on page 25 the momentum in unit volume of the medium at P is $N \times 4\pi\mu Nv \sin\theta$, or $4\pi\mu N^2 v \sin\theta$, and is in the direction of the component of the velocity of the Faraday tubes at right angles to their length. Now this is exactly the momentum which would be produced if the tubes were to carry with them, when they move at right angles to their length, a mass of the surrounding medium equal to $4\pi\mu N^2$ per unit volume, the tubes possessing no mass themselves and not carrying any of the medium with them when they glide through it parallel to their own length. We suppose in fact the tubes to behave very much as long and narrow cylinders behave when moving through water; these if moving endwise, *i.e.*, parallel to their length, carry very little water along with them, while when they move sideways, *i.e.*, at right angles to their axis, each unit length of the tube carries with it a finite mass of water. When the length of the cylinder is very great compared with its breadth, the mass of water carried by it when moving endwise may be neglected in comparison with that carried by it when moving sideways; if the tube had no mass beyond that which it possesses in virtue of the water it displaces, it would have mass for sideways but none for endwise motion.

We shall call the mass $4\pi\mu N^2$ carried by the tubes in unit volume the mass of the bound ether. It is a very suggestive fact that the electrostatic energy E in unit volume is proportional to M the mass of the bound ether in that volume. This can easily be proved as follows: $E = \frac{2\pi N^2}{K}$, where K is the specific inductive capacity of the medium; while $M = 4\pi\mu N^2$, thus,

$$E = \tfrac{1}{2}\frac{M}{\mu K};$$

but $\frac{1}{\mu K} = V^2$ where V is the velocity with which light travels through the medium, hence

$$E = \tfrac{1}{2} MV^2;$$

thus E is equal to the kinetic energy possessed by the bound mass when moving with the velocity of light.

The mass of the bound ether in unit volume is $4\pi\mu N^2$ where N is the number of Faraday tubes; thus, the amount of bound mass per unit length of each Faraday tube is $4\pi\mu N$. We have seen that this is proportional to the tension in each tube, so that we may regard the Faraday tubes as tightly stretched strings of variable mass and tension; the tension being, however, always proportional to the mass per unit length of the string.

Since the mass of ether imprisoned by a Faraday tube is proportional to N the number of Faraday tubes in unit volume, we see that the mass and momentum of a Faraday tube depend not merely upon the configuration and velocity of the tube under consideration, but also upon the number and velocity of the Faraday tubes in its neighborhood. We have many analogies to this in the case of dynamical systems; thus, in the case of a number of cylinders with their axes parallel, moving about in an incompressible liquid, the momentum of any cylinder depends upon the positions and velocities of the cylinders in its neighborhood. The following hydro-dynamical system is one by which we may illustrate the fact that the bound mass is proportional to the square of the number of Faraday tubes per unit volume.

Suppose we have a cylindrical vortex column of strength m placed in a mass of liquid whose velocity, if not disturbed by the vortex column, would be constant both in magnitude and direction, and at right angles to the axis of the vortex column. The lines of flow in such a case are represented in Fig. 11, where A is the section of the vortex

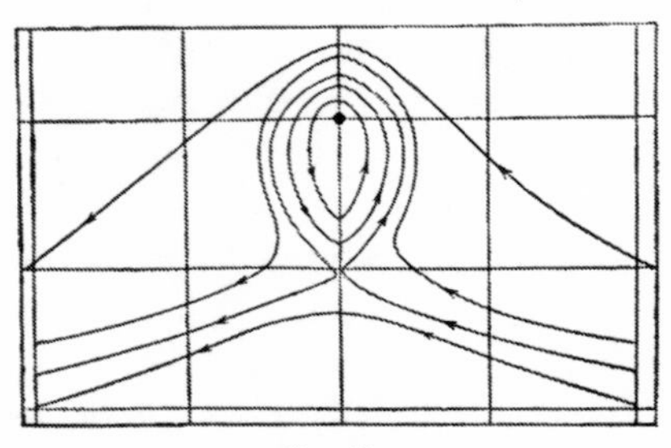

FIG. 11.

column whose axis is supposed to be at right angles to the plane of the paper. We see that some of these lines in the neighborhood of the column are closed curves. Since the liquid does not cross the lines of flow, the liquid inside a closed curve will always remain in the neighborhood of the column and will move with it. Thus, the column will imprison a mass of liquid equal to that enclosed by the largest of the closed lines of flow. If m is the strength of the vortex column and a the velocity of the undisturbed flow of the liquid, we can easily show that the mass of liquid imprisoned by the column is proportional to $\frac{m^2}{a^2}$. Thus, if we take m as proportional to the number of Faraday tubes in unit area, the system illustrates the connection between the bound mass and the strength of the electric field.

Effective of Velocity on the Bound Mass

I will now consider another consequence of the idea that the mass of a charged particle arises from the mass of ether bound by the Faraday tube associated with the charge. These tubes, when they move at right angles to their length, carry with them an appreciable portion of the ether through which they move, while when they move parallel to their length, they glide through the fluid without setting it in motion. Let us consider how a long, narrow cylinder, shaped like a Faraday tube, would behave when moving through a liquid.

Such a body, if free to twist in any direction, will not, as you might expect at first sight, move point foremost, but will, on the contrary, set itself broadside to the direction of motion, setting itself so as to carry with it as much of the fluid through which it is moving as possible. Many illustrations of this principle could be given, one very familiar one is that falling leaves do not fall edge first, but flutter down with their planes more or less horizontal.

If we apply this principle to the charged sphere, we see that the Faraday tubes attached to the sphere will tend to set themselves at right angles to the direction of motion of the sphere, so that if this principle were the only thing to be considered all the Faraday tubes would be forced up into the equatorial plane, *i.e.*, the plane at right angles to the direction of motion of the sphere, for in this position they would all be moving at right angles to their lengths. We must remember, however, that the Faraday tubes repel each other, so that if they were crowded into the equatorial region the pressure there would be greater than that near the pole. This would tend to thrust the Faraday tubes back into the position in which they are equally distributed all over the sphere. The actual distribution of the Faraday tubes is a compromise between these extremes. They are not all crowded into the equatorial plane, neither are they equally distributed, for they are more in the equatorial regions than in the others; the excess of the density of the tubes in these regions increasing with the speed with which the charge is moving.

When a Faraday tube is in the equatorial region it imprisons more of the ether than when it is near the poles, so that the displacement of the Faraday tubes from the pole to the equator will increase the amount of ether imprisoned by the tubes, and therefore the mass of the body.

It has been shown (see Heaviside, *Phil. Mag.*, April, 1889, "Recent Researches," p. 19) that the effect of the motion of the sphere is to displace each Faraday tube toward the equatorial plane, *i.e.*, the plane through the centre of the sphere at right angles to its direction of motion, in such a way that the projection of the tube on this plane remains the same as for the uniform distribution of tubes, but that the distance of every point in the tube from the equatorial plane is reduced in the proportion of $\sqrt{V^2-v^2}$ to V, where V is the velocity of light through the medium and v the velocity of the charged body.

From this result we see that it is only when the velocity of the charged body is comparable with the velocity of light that the change in distribution of the Faraday tubes due to the motion of the body becomes appreciable.

In "Recent Researches on Electricity and Magnetism," p. 21, I calculated the momentum I, in the space surrounding a sphere of radius a, having its centre at the moving charged body, and showed that the value of I is given by the following expression:

$$I = \frac{e^2}{2a} \frac{V^2}{(V^2 - v^2)^{\frac{1}{2}}} \left\{ \theta \left(1 - \tfrac{1}{4} \frac{V^2}{v^2}\right) + \tfrac{1}{2} \sin 2\theta \right.$$

$$\left. \left(1 + \tfrac{1}{4} \frac{V^2}{v^2} \cos 2\theta\right) \right\} ; \quad . \quad . \quad (1)$$

where as before v and V are respectively the velocities of the particle and the velocity of light, and θ is given by the equation

$$\sin \theta = \frac{v}{V}.$$

The mass of the sphere is increased in consequence of the charge by $\frac{I}{v}$, and thus we see from equation (1) that for velocities of the charged body comparable with that of light the mass of the body will increase with the velocity. It is evident from equation (1) that to detect the influence of velocity on mass we must use exceedingly small particles moving with very high velocities. Now, particles having masses far smaller than the mass of any known atom or molecule are shot out from radium with velocities approaching in some cases to that of light, and the ratio of the electric charge to the mass for particles of this kind has lately been made the subject of a very interesting investigation by Kaufmann, with the results shown in the following table; the first column contains the values of the velocities of the particle expressed in centimetres per second, the second column the value of the fraction $\frac{e}{m}$ where e is the charge and m the mass of the particle:

$v \times 10^{-10}$	$\frac{e}{m} \times 10^{-7}$
2.83	.62
2.72	.77
2.59	.975
2.48	1.17
2.36	1.31

We see from these values that the value of $\frac{e}{m}$ diminishes as the velocity increases, indicating, if we suppose the charge to remain constant, that the mass increases with the velocity. Kaufmann's results give us the means of comparing the part of the mass due to the electric charge with the part independent of the electrification; the second part of the mass is independent of the velocity. If then we find that the mass varies appreciably with the velocity, we infer that the part of the mass due to the charge must be appreciable in comparison with that independent of it. To calculate the effect of velocity on the mass of an electrified system we must make some assumption as to the nature of the system, for the effect on a charged sphere for example is not quite the same as that on a charged ellipsoid; but having made the assumption and calculated the theoretical effect of the velocity on the mass, it is easy to deduce the ratio of the part of the mass independent of the charge to that part which at any velocity depends upon the charge. Suppose that the part of the mass due to electrification is at a velocity v equal to $m_o f(v)$ where $f(v)$ is a known function of v, then if M_v, M_{v^1} are the observed masses at the velocities v and v^1 respectively and M the part of the mass independent of charge, then

$$M_v = M + m_o f(v),$$
$$M_{v^1} = M + m_o f(v^1),$$

two equations from which M and m_o can be determined. Kaufmann, on the assumption that the charged body behaved like a metal sphere, the distribution of the lines of force of which when moving has been determined by G. F. C. Searle, came to the conclusion that when the particle was moving slowly the "electrical mass" was about one-fourth of the whole mass. He was careful to point out that this fraction depends upon the assumption we make as to the nature of the moving body, as, for example, whether it is spherical or ellipsoidal, insulating or conducting; and that with other assumptions his experiments might show that the whole mass was electrical, which he evidently regarded as the most probable result.

In the present state of our knowledge of the constitution of matter, I do not think anything is gained by attributing to the small negatively charged bodies shot out by radium and other bodies the property of metallic conductivity, and I prefer the simpler assumption that the distribution of the lines of force round these particles is the same as that of the lines due to a charged point, provided we confine our attention to the field outside a small sphere of radius a having its centre at the charged point; on this supposition the part of the mass due to the charge is the value of $\frac{I}{v}$ in equation (1) on page 44. I have calculated from this expression the ratio of the masses of the rapidly moving particles given out by radium to the mass of the same particles when at rest, or moving slowly, on the assumption that *the whole of the mass is due to the charge* and have compared these results with the values of the same ratio as determined by Kaufmann's experiments. These results are given in Table (II), the first column of which contains the values of v, the velocities of the particles; the second ρ, the number of times the mass of a particle moving with this velocity exceeds the mass of the same particle when at rest, determined by equation (1); the third column ρ^1, the value of this quantity found by Kaufmann in his experiments.

TABLE II.

$v \times 10^{-10} \frac{cm}{sec}$	ρ	ρ^1
2.85	3.1	3.09
2.72	2.42	2.43
2.59	2.0	2.04
2.48	1.66	1.83
2.36	1.5	1.65

These results support the view that the *whole* mass of these electrified particles arises from their charge.

We have seen that if we regard the Faraday tubes associated with these moving particles as being those due to a moving point charge, and confine our attention to the part of the field which is outside a sphere of radius a concentric with the charge, the mass m due to the charge e on the particle is, when the particle is moving slowly, given by the equation $m = \frac{2}{3}\frac{\mu e^2}{a}$.

In a subsequent lecture I will explain how the values of $m\,e$ and e have been determined; the result of these determinations is that $\frac{m}{e} = 10^{-7}$ and $e = 1.2 \times 10^{-20}$ in *C. G. S.* electrostatic units. Substituting these values in the expression for m we find that a is about 5×10^{-14} *cm*, a length very small in comparison with the value 10^{-8} *cm*, which is usually taken as a good approximation to the dimensions of a molecule.

We have regarded the mass in this case as due to the mass of ether carried along by the Faraday tubes associated with the charge. As these tubes stretch out to an infinite distance, the mass of the particle is as it were diffused through space, and has no definite limit. In consequence, however, of the very small size of the particle and the fact that the mass of ether carried by the tubes (being proportional to the square of the density of the Faraday tubes) varies inversely as the fourth power of the distance from the particle, we find by a simple calculation that all but the most insignificant fraction of mass is confined to a distance from the particle which is very small indeed compared with the dimensions ordinarily ascribed to atoms.

In any system containing electrified bodies a part of the mass of the system will consist of the mass of the ether carried along by the Faraday tubes associated with the electrification. Now one view of the constitution of matter—a view, I hope to discuss in a later lecture—is that the atoms of the various elements are collections of positive and negative charges held together mainly by their electric attractions, and, moreover, that the negatively electrified particles in the atom (corpuscles I have termed them) are identical with those small negatively electrified particles whose properties we have been discussing. On this view of the constitution of matter, part of the mass of any body would be the mass of the ether dragged along by the Faraday tubes stretching across the atom between the positively and negatively electrified constituents. The view I wish to put before you is that it is not merely a part of the mass of a body which arises in this way, but that the *whole* mass of any body is just the mass of ether surrounding the body which is carried along by the Faraday tubes associated with the atoms of the body. In fact, that all mass is mass of the ether, all momentum, momentum of the ether, and all kinetic energy, kinetic energy of the ether. This view, it should be said, requires the density of the ether to be immensely greater than that of any known substance.

It might be objected that since the mass has to be carried along by the Faraday tubes and since the disposition of these depends upon the relative position of the electrified bodies, the mass of a collection of a numberof positively and negatively electrified bodies would be constantly changing with the positions of these bodies, and thus that mass instead of being, as observation and experiment have shown, constant to a very high degree of approximation, should vary with changes in the physical or chemical state of the body.

These objections do not, however, apply to such a case as that contemplated in the preceding theory, where the dimensions of one set of the electrified bodies—the negative ones—are excessively small in comparison with the distances separating the various members of the system of electrified bodies.

When this is the case the concentration of the lines of force on the small negative bodies—the corpuscles—is so great that practically the whole of the bound ether is localized around these bodies, the amount depending only on their size and charge. Thus, unless we alter the number or character of the corpuscles, the changes occurring in the mass through any alteration in their relative positions will be quite insignificant in comparison with the mass of the body.

CHAPTER III

EFFECTS DUE TO ACCELERATION OF THE FARADAY TUBES

Röntgen Rays and Light

WE have considered the behavior of the lines of force when at rest and when moving uniformly, we shall in this chapter consider the phenomena which result when the state of motion of the lines is changing.

Let us begin with the case of a moving charged point, moving so slowly that the lines of force are uniformly distributed around it, and consider what must happen if we suddenly stop the point. The Faraday tubes associated with the sphere have inertia; they are also in a state of tension, the tension at any point being proportional to the mass per unit length. Any disturbance communicated to one end of the tube will therefore travel along it with a constant and finite velocity; the tube in fact having very considerable analogy with a stretched string. Suppose we have a tightly stretched vertical string moving uniformly, from right to left, and that we suddenly stop one end, A, what will happen to the string? The end A will come to rest at once, but the forces called into play travel at a finite rate, and each part of the string will in virtue of its inertia continue to move as if nothing had happened to the end A until the disturbance starting from A reaches it. Thus, if V is the velocity with which a disturbance travels along the string, then when a time, t, has elapsed after the stoppage of A, the parts of the string at a greater distance than Vt from A will be unaffected by the stoppage, and will have the position and velocity they would have had if the string had continued to move uniformly forward. The shape of the string at successive intervals will be as shown in Fig. 12, the length of

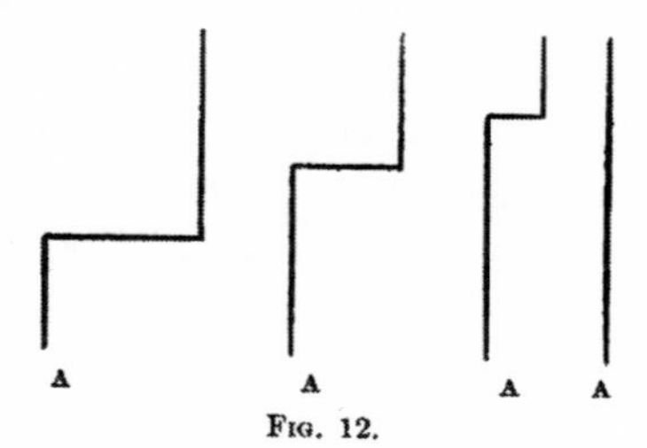

Fig. 12.

the horizontal portion increasing as its distance from the fixed end increases.

Let us now return to the case of the moving charged particle which we shall suppose suddenly brought to rest, the time occupied by the stoppage being τ. To find the configuration of the Faraday tubes after a time t has elapsed since the beginning of the process of bringing the charged particle to rest, describe with the charged particle as centre two spheres, one having the radius Vt, the other the radius $V(t-\tau)$, then, since no disturbance can have reached the Faraday tubes situated outside the outer sphere, these tubes will be in the position they would have occupied if they had moved forward with the velocity they possessed at the moment the particle was stopped, while inside the inner sphere, since the disturbance has passed over the tubes, they will be in their final positions. Thus, consider a tube which, when the particle was stopped was along the line OPQ (Fig. 13); this will be the final position of the tube; hence at the time t the portion of this tube inside the inner sphere will occupy the position OP, while the portion $P'Q'$ outside the outer sphere will be in the position it would have occupied if the particle had not been reduced to rest, *i.e.*, if O' is the position the particle would have occupied if it had not been stopped, $P'Q'$ will be a straight line passing through O'. Thus, to preserve its continuity the tube must bend round in the shell between the two spheres, and thus be distorted into the shape $OPP'Q'$. Thus, the tube which before the stop-

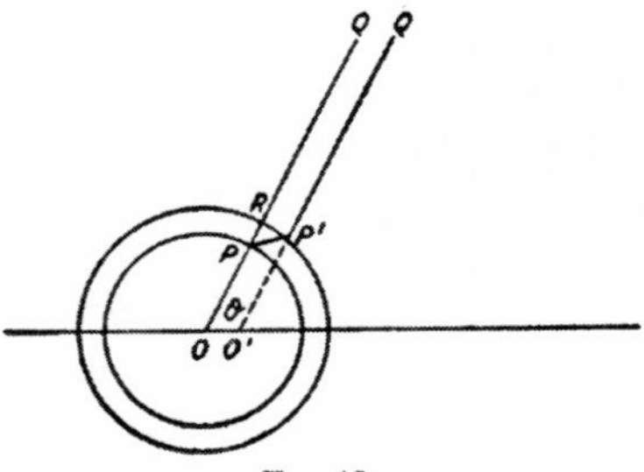

FIG. 13.

page of the particle was radial, has now in the shell a tangential component, and this tangential component implies a tangential electric force. The stoppage of the particle thus produces a radical change in the electric field due to the particle, and gives rise, as the following calculation will show, to electric and magnetic forces much greater than those existing in the field when the particle was moving steadily.

If we suppose that the thickness δ of the shell is so small that the portion of the Faraday tube inside it may be regarded as straight, then if T is the tangential electric force inside the pulse, R the radial force, we have

$$\frac{T}{R} = \frac{P'R}{PR} = \frac{00' \sin\theta}{\delta} = \frac{vt\sin\theta}{\delta}. \quad (1)$$

Where v is the velocity with which the particle was moving before it was stopped, θ the angle OP makes with the direction of motion of the particle, t the time which has elapsed since the particle was stopped; since $R = \frac{e}{OP^2}$ and $OP = Vt$ where V is the velocity of light, we have, if $r = OP$,

$$T = \frac{ev}{V}\frac{\sin\theta}{r\delta}. \quad (2)$$

The tangential Faraday tubes moving forward with the velocity V will produce at P a magnetic force H equal to VT, this force will be at right angles to the plane of the paper and in the opposite direction to the magnetic force existing at P before the stoppage of the particle; since its magnitude is given by the equation,

$$H = \frac{ev\sin\theta}{r\delta},$$

it exceeds the magnetic force $\frac{ev\sin\theta}{\gamma^2}$ previously existing in the proportion of r to δ. Thus, the pulse produced by the stoppage of the particle is the seat of intense electric and magnetic forces which diminish inversely as the distance from the charged particle, whereas the forces before the particle was stopped diminished inversely as the square of the distance; this pulse travelling outward with the velocity of light constitutes in my opinion the Röntgen rays which are produced when the negatively electrified particles which form the cathode rays are suddenly stopped by striking against a solid obstacle.

The energy in the pulse can easily be shown to be equal to

$$\frac{2}{3}\frac{e^2v^2}{\delta},$$

this energy is radiated outward into space. The amount of energy thus radiated depends upon δ, the thickness of the pulse, *i.e.*, upon the abruptness with which the particle is stopped; if the particle is stopped instantaneously the whole energy in the field will be absorbed in the pulse and radiated away, if it is stopped gradually only a fraction of the energy will be radiated into space, the remainder will appear as heat at the place where the cathode rays were stopped.

It is easy to show that the momentum in the

pulse at any instant is equal and opposite to the momentum in the field outside the pulse; as there is no momentum in the space through which the pulse has passed, the whole momentum in the field after the particle is stopped is zero.

The preceding investigation only applies to the case when the particle was moving so slowly that the Faraday tubes before the stoppage of the pulse were uniformly distributed; the same principles, however, will give us the effect of stopping a charged particle whenever the distribution of the Faraday tubes in the state of steady motion has been determined.

Let us take, for example, the case when the particle was initially moving with the velocity of light; the rule stated on page 43 shows that before the stoppage the Faraday tubes were all congregated in the equatorial plane of the moving particle. To find the configuration of the Faraday tubes after a time t we proceed as before by finding the configuration at that time of the tubes, if the particle had not been stopped. The tubes would in that case have been in a plane at a distance Vt in front of the particle. Draw two spheres having their centres at the particle and having radii respectively equal to Vt and $V(t-\tau)$, where τ is the time occupied in stopping the particle; outside the outer sphere the configuration of the tubes will be the same as if the particle had not been stopped, *i. e.*, the tubes will be the plane at the distance Vt in front of the particle, and this plane will touch the outer sphere. Inside the inner sphere the Faraday tubes will be uniformly distributed, hence to preserve continuity these tubes must run round in the shell to join the sphere as in Fig. 14. We thus have in this case two pulses, one a plane pulse propagated in the direction in which the particle was moving before it was stopped, the other a spherical pulse travelling outward in all directions.

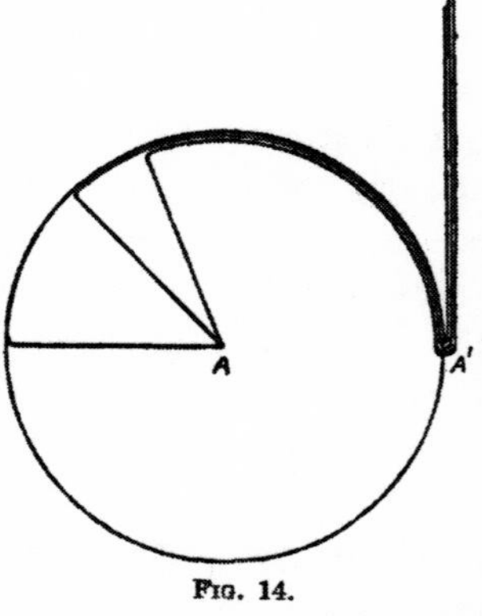

FIG. 14.

The preceding method can be applied to the case when the charged particle, instead of being stopped, has its velocity altered in any way; thus, if the velocity v of the particle instead of being reduced to zero is merely diminished by Δv, we can show, as on page 57, that it will give rise to a pulse in which the magnetic force H is given by the equation

$$H = \frac{e \Delta v \sin \theta}{r \delta},$$

and the tangential electric force T by

$$T = \frac{e . \Delta v \sin \theta}{V r . \delta}.$$

Now the thickness δ of the pulse is the space passed over by a wave of light during the time the velocity of the particle is changing, hence if δt is the time required to produce the change Δv in the velocity $\delta = V \delta t$, hence we have

$$H = \frac{e}{V} \frac{\Delta v}{\delta t} \frac{\sin \theta}{r} \qquad T = \frac{e}{V^2} \frac{\Delta v}{\delta t} \frac{\sin \theta}{r};$$

but $\frac{\Delta v}{\delta t}$ is equal to $-f$, where f is the acceleration of the particle, hence we have

$$H = -\frac{e}{V} f \frac{\sin \theta}{r} \qquad T = -\frac{e}{V^2} f \frac{\sin \theta}{r}.$$

These equations show that a charged particle whose motion is being accelerated produces a pulse of electric and magnetic forces in which the forces vary inversely as the distance from the particle.

Thus, if a charged body were made to vibrate in

such a way that its acceleration went through periodic changes, periodic waves of electric and magnetic force would travel out from the charged body. These waves would, on the Electromagnetic Theory of Light, be light waves, provided the periodic changes in the acceleration of the charged body took place with sufficient rapidity. The method we have been investigating, in which we consider the effect produced on the configuration of the Faraday tubes by changes in the motion of the body, affords a very simple way of picturing to ourselves the processes going on during the propagation of a wave of light through the ether. We have regarded these as arising from the propagation of transverse tremors along the tightly stretched Faraday tubes; in fact, we are led to take the same view of the propagation of light as the following extracts from the paper, "Thoughts on Ray Vibrations," show to have been taken by Faraday himself. Faraday says, "The view which I am so bold to put forward considers therefore radiations as a high species of vibration in the lines of force which are known to connect particles and also masses together."

This view of light as due to the tremors in tightly stretched Faraday tubes raises a question which I have not seen noticed. The Faraday tubes stretching through the ether cannot be regarded as entirely filling it. They are rather to be looked upon as discrete threads embedded in a continuous ether, giving to the latter a fibrous structure; but if this is the case, then on the view we have taken of a wave of light the wave itself must have a structure, and the front of the wave, instead of being, as it were, uniformly illuminated, will be represented by a series of bright specks on a dark ground, the bright specks corresponding to the places where the Faraday tubes cut the wave front.

Such a view of the constitution of a light wave would explain a phenomenon which has always struck me as being very remarkable and difficult to reconcile with the view that a light wave, or rather in this case a Röntgen ray, does not possess a structure. We have seen that the method of propagation and constitution of a Röntgen ray is the same as in a light wave, so that any general consideration about structure in Röntgen rays will apply also to light waves. The phenomenon in question is this: Röntgen rays are able to pass very long distances through gases, and as they pass through the gas they ionize it, splitting up the molecules into positive and negative ions; the number of molecules so split up is, however, an exceedingly small fraction, less than one-billionth, even for strong rays, of the number of molecules in the gas. Now, if the conditions in the front of the wave are uniform, all the molecules of the gas are exposed to the same conditions; how is it then that so small a proportion of them are split up? It might be argued that those split up are in some special condition—that they possess, for example, an amount of kinetic energy so much exceeding the average kinetic energy of the molecules of the gas that, in accordance with Maxwell's Law of Distribution of Kinetic energy, their number would be exceedingly small in comparison with the whole number of molecules of the gas; but if this were the case the same law of distribution shows that the number in this abnormal condition would increase very rapidly with the temperature, so that the ionization produced by the Röntgen rays ought to increase very rapidly as the temperature increases. Recent experiments made by Mr. McClung in the Cavendish Laboratory show that no appreciable increase in the ionization is produced by raising the temperature of a gas from 15° C. to 200° C., whereas the number of molecules possessing an abnormal amount of kinetic energy would be enormously increased by this rise in temperature. The difficulty in explaining the small ionization is removed if, instead of supposing the front of the Röntgen ray to be uniform, we suppose that it consists of specks of great intensity separated by considerable intervals where the intensity is very small, for in this case all the molecules in the field, and probably even different parts of the same molecule, are not exposed to the same conditions, and the case becomes analogous to a swarm of cathode rays passing through the gas, in which case the number of molecules which get into collision with the rays may be a very small fraction of the whole number of molecules.

To return, however, to the case of the charged particle whose motion is accelerated, we have seen that from the particle electric and magnetic forces start and travel out radially with the velocity of light, both the radial and magnetic forces being at right angles to the direction in which they are travelling; but since (see page 25) each unit volume of the electro-magnetic field has an amount of momentum equal to the product of the density of the Faraday tube and the magnetic force, the direction of the momentum being at right angles to both these quantities, there will be the wave due to the acceleration of the charged particle, and indeed in any electric or light wave momentum in the direction of propagation of the wave. Thus, if any such wave, for example a wave of light, is absorbed by the substance through which it is passing, the momentum in the wave will be communicated to the absorbing substance, which will, therefore, experience a force tending to push it in the direction the light is travelling. Thus, when light falls normally on a blackened absorbing substance, it will repel that substance. This repulsion resulting from radiation was shown by Maxwell to be a consequence of the Electromagnetic Theory of Light; it has lately been detected and measured by Lebedew by some most beautiful experiments, which have been confirmed and extended by Nichols and Hull.

The pressure experienced by the absorbing substance will be proportional to its area, while the weight of the substance is proportional to its volume. Thus, if we halve the linear dimensions we reduce the weight to one-eighth while we only reduce the pressure of radiation to one-quarter; thus, by sufficiently reducing the size of the absorbing body we must arrive at a stage when the forces due to radiation exceed those which, like weight, are proportional to the volume of the substance. On this principle, knowing the intensity of the radiation from the sun, Arrhenius has shown that for an opaque sphere of unit density 10^{-5} cm. in diameter the repulsion due to the radiation from the sun would just balance the sun's attraction, while all bodies smaller than this would be repelled from the sun, and he has applied this principle to explain the phenomena connected with the tails of comets. Poynting has recently shown that if two spheres of unit density about 39 cm. in diameter are at the temperature of 27° C. and protected from all external radiation, the repulsion due to the radiation emitted from the spheres will overpower their gravitational attraction so that the spheres will repel each other.

Again, when light is refracted and reflected at a transparent surface, the course of the light and therefore the direction of momentum is changed, so that the refracting substance must have momentum communicated to it. It is easy to show that even when the incidence of the light is oblique the momentum communicated to the substance is normal to the refracting surface. There are many interesting problems connected with the forces experienced by refracting prisms when light is passing through them which will suggest themselves to you if you consider the changes in momentum experienced by the light wave in its course through the prism. Tangential forces due to light have not, so far as I know, been detected experimentally. These, however, must exist in certain cases; such, for example, as when light incident obliquely is imperfectly reflected from a metallic surface.

The waves of electric and magnetic force which radiate from an accelerated charge particle carry energy with them. This energy is radiated into space, so that the particle is constantly losing energy. The rate at which energy is radiating from the particle can easily be shown to be $\frac{1}{3}\frac{e^2f^2}{v}$ where e is the charge on the particle, f its acceleration, and V the velocity of light. If we take into account this loss of energy by the particle when its motion is being accelerated, we find some interesting results. Thus, for example, if a particle of mass m and charge e starting from rest is acted upon by a constant electric force, X, the particle does not at once attain the acceleration $\frac{Xe}{m}$ as it would if there were no loss of energy by radiation; on the contrary, the acceleration of the particle is initially zero, and it is not until after the lapse of a time comparable with $\frac{e^2}{Vm}$ that the particle acquires even an appreciable fraction of its final acceleration. Thus, the rate at which the particle loses energy is during the time $\frac{e^2}{Vm}$ very small compared with the ultimate rate. Thus, if the particle were acted on by a wave of electric force which only took a time comparable with $\frac{e^2}{Vm}$ to pass over the particle, the amount of energy radiated by the particle would be a very much smaller fraction of the energy in the wave than it would be if the particle took a time equal to a considerable multiple of $\frac{e^2}{Vm}$ to pass over the particle. This has an important application in explaining the greater penetrating power of "hard" Röntgen rays than of "soft" ones. The "hard" rays correspond to thin pulses, the "soft" ones to thick ones; so that a smaller proportion of the energy in the "hard" rays will be radiated away by the charged particles over which they pass than in the case of the "soft" rays.

By applying the law that the rate at which energy is radiating is equal to $\frac{1}{3}\frac{e^2f^2}{V}$ to the case of a charged particle revolving in a circular orbit under an attractive force varying inversely as the square of the distance, we find that in this case the rate of radiation is proportional to the eighth power of the velocity, or to the fourth power of the energy. Thus, the rate of loss of energy by radiation increases very much more rapidly than the energy of the moving body.

CHAPTER IV

THE ATOMIC STRUCTURE OF ELECTRICITY

HITHERTO we have been dealing chiefly with the properties of the lines of force, with their tension, the mass of ether they carry along with them, and with the propagation of electric disturbances along them; in this chapter we shall discuss the nature of the charges of electricity which form the beginnings and ends of these lines. We shall show that there are strong reasons for supposing that these changes have what may be called an atomic structure; any charge being built up of a number of finite individual charges, all equal to each other: just as on the atomic theory of matter a quantity of hydrogen is built up of a number of small particles called atoms, all the atoms being equal to each other. If this view of the structure of electricity is correct, each extremity of a Faraday tube will be the place from which a constant fixed number of tubes start or at which they arrive.

Let us first consider the evidence given by the laws of the electrolysis of liquids. Faraday showed that when electricity passes through a liquid electrolyte, the amount of negative electricity given up to the positive electrode, and of positive electricity given to the negative electrode, is proportional to the number of atoms coming up to the electrode. Let us first consider monovalent elements, such as hydrogen, chlorine, sodium, and so on; he showed that when the same number of atoms of these substances deliver up their charges to the electrode, the quantity of electricity communicated is the same whether the carriers are atoms of hydrogen, chlorine, or sodium, indicating that each atom of these elements carries the same charge of electricity. Let us now go to the divalent elements. We find again that the ions of all divalent elements carry the same charge, but that a number of ions of the divalent element carry twice the charge carried by the same number of ions of a univalent element, showing that each ion of a divalent element carries twice as much charge as the univalent ion; again, a trivalent ion carries three times the charge of a univalent ion, and so on. Thus, in the case of the electrolysis of solutions the charges carried by the ions are either the charge on the hydrogen ion or twice that charge, or three times the charge, and so on. The charges we meet with are always an integral multiple of the charge carried by the hydrogen atom; we never meet with fractional parts of this charge. This very remarkable fact shows, as Helmholtz said in the Faraday lecture, that "if we accept the hypothesis that the elementary substances are composed of atoms, we cannot avoid the conclusion that electricity, positive as well as negative, is divided into definite elementary portions which behave like atoms of electricity."

When we consider the conduction of electricity through gases, the evidence in favor of the atomic character of electricity is even stronger than it is in the case of conduction through liquids, chiefly because we know more about the passage of electricity through gases than through liquids.

Let us consider for a moment a few of the properties of gaseous conduction. When a gas has been put into the conducting state—say, by exposure to Röntgen rays—it remains in this state for a sufficiently long time after the rays have ceased to enable us to study its properties. We find that we can filter the conductivity out of the gas by sending the gas through a plug of cotton-wool, or through a water-trap. Thus, the conductivity is due to something mixed with the gas which can be filtered out of it; again, the conductivity is taken out of the gas when it is sent through a strong electric field. This result shows that the constituent to which the conductivity of the gas is due consists of charged particles, the conductivity arising from the motion of these particles in the electric field. We have at the Cavendish Laboratory measured the charge of electricity carried by those particles.

The principle of the method first used is as follows. If at any time there are in the gas n of these particles charged positively and n charged negatively, and if each of these carries an electric charge e, we can easily by electrical methods determine $n\ e$, the quantity of electricity of our sign present in the gas. One method by which this can be done is to enclose the gas between two parallel metal plates, one of which is insulated. Now suppose we suddenly charge up the other plate positively to a very high potential, this plate will now repel the positive particles in the gas, and these before they have time to combine with the negative particles will be driven against the insulated plate. Thus, all the positive charge in the gas will be driven against the insulated plate, where it can be measured by an electrometer. As this charge is equal to $n\ e$ we can in this way easily determine $n\ e$: if then we can devise a means of measuring n we shall be able to find e. The method by which I determined n was founded on the discovery by C. T. R. Wilson that the charged particles act as nuclei round which small drops of water condense, when the particles are surrounded by damp air cooled below the saturation point. In dust-free air, as Aitken showed, it is very difficult to get a fog when damp air is cooled, since there are no nuclei for the drops to condense around; if there are charged particles in the dust-free air, however, a fog will be deposited round these by a supersaturation far less than that required to produce any appreciable effect when no charged particles are present.

Thus, in sufficiently supersaturated damp air a cloud is deposited on these charged particles, and they are thus rendered visible. This is the first step toward counting them. The drops are, however, far too small and too numerous to be counted directly. We can, however, get their number indirectly as follows: suppose we have a number of these particles in dust-free air in a closed vessel, the air being saturated with water vapor, suppose now that we produce a sudden expansion of the air in the vessel; this will cool the air, it will be supersaturated with vapor, and drops will be deposited round the charged particles. Now if we know the amount of expansion produced we can calculate the cooling of the gas, and therefore the amount of water deposited. Thus, we know the volume of water in the form of drops, so that if we know the volume of one drop we can deduce the number of drops. To find the size of a drop we make use of an investigation by Sir George Stokes on the rate at which small spheres fall through the air. In consequence of the viscosity of the air small bodies fall exceedingly slowly, and the smaller they are the slower they fall. Stokes showed that if a is the radius of a drop of water, the velocity v with which it falls through the air is given by the equation

$$v = \frac{2}{9} \frac{g a^2}{\mu};$$

when g is the acceleration due to gravity $= 981$ and μ the coefficient of viscosity of air $= .00018$; thus

$$v = 1.21 \times 10^6 a^2;$$

hence if we can determine v we can determine the radius and hence the volume of the drop.

But v is evidently the velocity with which the cloud round the charged particle settles down, and can easily be measured by observing the movement of the top of the cloud. In this way I found the volume of the drops, and thence n the number of particles. As $n\,e$ had been determined by electrical measurements, the value of e could be deduced when n was known; in this way I found that its value is

$$3.4 \times 10^{-10} \text{ Electrostatic C. G. S. units.}$$

Experiments were made with air, hydrogen, and carbonic acid, and it was found that the ions had the same charge in all these gases; a strong argument in favor of the atomic character of electricity.

We can compare the charge on the gaseous ion with that carried by the hydrogen ion in the electrolysis of solutions in the following way: We know that the passage of one electro-magnetic unit of electric charge, or 3×10^{10} electrostatic units, through acidulated water liberates 1.23 c.c. of hydrogen at the temperature 15° C. and pressure of one atmosphere; if there are N molecules in a c.c. of a gas at this temperature and pressure the number of hydrogen ions in 1.23 c.c. is 2.46 N, so that if E is the charge on the hydrogen ion in the electrolyis of solution,

$$2.46 \, NE = 3 \times 10^{10},$$
$$\text{or} \quad E = 1.22 \times 10^{10} \div N.$$

Now, e, the charge on the gas ion is 3.4×10^{-10}, hence if $N = 3.6 \times 10^{19}$ the charge on the gaseous ion will equal the charge on the electrolytic ion. Now, in the kinetic theory of gases methods are investigated for determining this quantity N, or Avogadro's Constant, as it is sometimes called; the values obtained by this theory vary somewhat with the assumptions made as to the nature of the molecule and the nature of the forces which one molecule exerts on another in its near neighborhood. The value 3.6×10^{19} is, however, in good agreement with some of the best of these determinations, and hence we conclude that the charge on the gaseous ion is equal to the charge on the electrolytic ion.

Dr. H. A. Wilson, of the Cavendish Laboratory, by quite a different method, obtained practically the same value for e as that given above. His method was founded on the discovery by C. T. R. Wilson that it requires less supersaturation to deposit clouds from moist air on negative ions than it does on positive. Thus, by suitably choosing the supersaturation, we can get the cloud deposited on the negative ions alone, so that each drop in the cloud is negatively charged; by observing the rate at which the cloud falls we can, as explained above, determine the weight of each drop. Now, suppose we place above the cloud a positively electrified plate, the plate will attract the cloud, and we can adjust the charge on the plate until the electric attraction just balances the weight of a drop, and the drops, like Mahomet's coffin, hang stationary in the air; if X is the electric force then the electric attraction on the drop is Xe, when e is the charge on the drop. As $X e$ is equal to the weight of the drop which is known, and as we can measure X, e can be at once determined.

Townsend showed that the charge on the gaseous ion is equal to that on the ion of hydrogen in ordinary electrolysis, by measuring the coefficient of diffusion of the gaseous ions and comparing it with the velocity acquired by the ion under a given electric force. Let us consider the case of a volume of ionized gas between two horizontal planes, and suppose that as long as we keep in any horizontal layer the number of ions remains the same, but that the number varies as we pass from one layer to another; let x be the distance of a layer from the lower plane, n the number of ions of one sign in unit volume of this layer, then if D be the coefficient of diffusion of the ions, the number of ions which in one second pass downward through unit area of the layer is

$$D\frac{dn}{dx};$$

so that the average velocity of the particles downward is

$$\frac{D}{n}\frac{dn}{dx}.$$

The force which sets the ions in motion is the variation in the partial pressure due to the ions; if this pressure is equal to p, the force acting on the ions in a unit volume is $\frac{dp}{dx}$, and the average force per ion is $\frac{1}{n}\frac{dp}{dx}$. Now we can find the velocity which an ion acquires when acted upon by a known force by measuring, as Rutherford and Zeleny have done, the velocities acquired by the ions in an electric field. They showed that this velocity is proportional to the force acting on the ion, so that if A is the velocity when the electric force is X and when the force acting on the ion is therefore $X e$, the velocity for unit force will be $\frac{A}{X e}$, and the velocity when the force is $\frac{1}{n} \frac{dp}{dx}$ will therefore be

$$\frac{1}{n} \frac{dp}{dx} \frac{A}{Xe};$$

this velocity we have seen, however, to be equal to

$$\frac{D}{n} \frac{dn}{dx};$$

hence we have

$$\frac{dp}{dx} \frac{A}{Xe} = D \frac{dn}{dx}. \qquad (1)$$

Now if the ions behave like a perfect gas, the pressure p bears a constant ratio to n, the number of ions per unit volume. This ratio is the same for all gases, at the same temperature, so that if N is Avogadro's constant, *i.e.*, the number of molecules in a cubic centimetre of gas at the atmospheric pressure P

$$\frac{p}{P} = \frac{n}{N},$$

and equation (1) gives us

$$\frac{PA}{XD} = Ne.$$

Thus, by knowing D and a we can find the value of Ne. In this way Townsend found that Ne was the same in air, hydrogen, oxygen, and carbonic acid, and the mean of his values was $Ne = 1.24 \times 10^{10}$. We have seen that if E is the charge on the hydrogen ion

$$NE = 1.22 \times 10^{10}.$$

Thus, these experiments show that $e = E$, or that the charge on the gaseous ion is equal to the charge carried by the hydrogen ion in the electrolysis of solutions.

The equality of these charges has also been proved in a very simple way by H. A. Wilson, who introduced per second into a volume of air at a very high temperature, a measured quantity of the vapor of metallic salts. This vapor got ionized and the mixture of air and vapor acquired very considerable conductivity. The current through the vapor increased at first with the electromotive force used to drive it through the gas, but this increase did not go on indefinitely, for after the current had reached a certain value no further increase in the electromotive force produced any change in the current. The current, as in all cases of conduction through gases, attained a maximum value called the "saturation current," which was not exceeded until the electric field applied to the gas approached the intensity at which sparks began to pass through the gas. Wilson found that the saturation current through the salt vapor was just equal to the current which if it passed through an aqueous solution of the salt would electrolyse in one second the same amount of salt as was fed per second into the hot air.

It is worth pointing out that this result gives us a method of determining Avogadro's Constant which is independent of any hypothesis as to the shape or size of molecules, or of the way in which they act upon each other. If N is this constant, e the charge on an ion, then $N e = 1.22 \times 10^{10}$ and we have seen that $e = 3.4 \times 10^{10}$, so that $N = 3.9 \times 10^{19}$.

Thus, whether we study the conduction of electricity through liquids or through gases, we are led to the conception of a natural unit or atom of electricity of which all charges are integral multiples, just as the mass of a quantity of hydrogen is an integral multiple of the mass of a hydrogen atom.

Mass of the Carriers of Electricity

We must now pass on to consider the nature of the systems which carry the charges, and in order to have the conditions as simple as possible let us begin with the case of a gas at a very low pressure, where the motion of the particles is not impeded by collisions with the molecules of the gas. Let us suppose that we have a particle of mass m, carrying a charge e, moving in the plane of the paper, and that it is acted on by a uniform magnetic field at right angles to this plane. We have seen that under these circumstances the particle will be acted upon by a mechanical force equal to $H e v$, where H is the magnetic force and v the velocity of the particle. The direction of this force is in the plane of the paper at right angles to the path of the particle. Since the force is always at right angles to the direction of motion of the particle, the velocity of the particle and therefore the magnitude of the force acting upon it will not alter, so that the path of the particle will be that described by a body acted upon by a constant normal force. It is easy to show that this path is a circle whose radius a is given by the equation

$$a = \frac{m v}{e H}. \qquad (1)$$

The velocity v of the particle may be determined by the following method: Suppose the particle is moving horizontally in the plane of the paper, through a uniform magnetic field H, at right angles to this plane, the particle will be acted upon by a vertical force equal to $H e v$. Now, if in addition to the magnetic force we apply a vertical electric force X, this will exert a vertical mechanical force $X e$ on the moving particle. Let us arrange the direction of X so that this force is in the opposite direction to that due to the magnet, and adjust the value of X until the two forces are equal. We can tell when this adjustment has been made, since in this case the motion of the particle under the action of the electric and magnetic forces will be the same as when both these forces are absent. When the two forces are equal we have

$$X e = H e v, \quad \text{or}$$

$$v = \frac{X}{H}. \qquad (2)$$

Hence if we have methods of tracing the motion of the particle, we can measure the radius a of the circle into which it is bent by a constant magnetic force, and determine the value of the electric force required to counteract the effect of the magnetic force. Equations (1) and (2) then give us the means of finding both v and $\frac{e}{m}$.

Values of $\frac{e}{m}$ for Negatively Electrified Particles in Gases at Low Pressures

The value of $\frac{e}{m}$ has been determined in this way for the negatively electrified particles which form the cathode rays which are so conspicuous a part of the electric discharge through a gas at low pressures; and also for the negatively electrified particles emitted by metals, (1) when exposed to ultra-violet light, (2) when raised to the temperature of incandescence. These experiments have led to the very remarkable result that the value of $\frac{e}{m}$ is the same whatever the nature of the gas in which the particle may be found, or whatever the nature of the metal from which it may be supposed to have proceeded. In fact, in every case in which the value of $\frac{e}{m}$ has been determined for negatively electrified particles moving with velocities considerably less than the velocity of light, it has been found to have the constant value about 10^7, the units being the centimetre, gram, and second, and the charge being measured in electromagnetic units. As the value of $\frac{e}{m}$ for the hydrogen ion in the electrolysis of liquids is only 10^4, and as we have seen the charge on the gaseous ion is equal to that on the hydrogen ion in ordinary electrolysis, we see that the mass of a carrier of the negative charge must be only about one-thousandth part of the mass of hydrogen atom; the mass was for a long time regarded as the smallest mass able to have an independent existence.

I have proposed the name corpuscle for these units of negative electricity. These corpuscles are the same however the electrification may have arisen or wherever they may be found. Negative electricity, in a gas at a low pressure, has thus a structure analogous to that of a gas, the corpuscles taking the place of the molecules. The "negative electric fluid," to use the old notation, resembles a gaseous fluid with a corpuscular instead of a molecular structure.

Carriers of Positive Electrification

We can apply the same methods to determine the values of $\frac{e}{m}$ for the carriers of positive electrification. This has been done by Wien for the positive electrification found in certain parts of the discharge in a vacuum tube, and I have measured $\frac{e}{m}$ for the positive electrification given off by a hot wire. The results of these measurements form a great contrast to those for the negative electrification, for $\frac{e}{m}$ for the positive charge, instead of having, as it has for the negative, the constant high value 10^7, is found never to have a value greater than 10^4, the value it would have if the carrier were the atom of hydrogen. In many cases the value of $\frac{e}{m}$ is very much less than 10^4, indicating that in these cases the positive charge is carried by atoms having a greater mass than that of the hydrogen atom. The value of $\frac{e}{m}$ varies with the nature of the electrodes and with the gas in the discharge tube, just as it would if the carriers of the positive charge were the atoms of the elements which happened to be present when the positive electrification was produced.

These results lead us to a view of electrification which has a striking resemblance to that of Franklin's "One Fluid Theory of Electricity." Instead of taking, as Franklin did, the electric fluid to be positive electricity we take it to be negative. The "electric fluid" of Franklin corresponds to an assemblage of corpuscles, negative electrification being a collection of these corpuscles. The transference of electrification from one place to another is effected by the motion of corpuscles from the place where is a gain of positive electrification to the place where there is a gain of negative. A positively electrified body is one that has lost some of its corpuscles. We have seen that the mass and charge of the corpuscles have been determined directly by experiment. We in fact know more about the "electric fluid" than we know about such fluids as air or water.

CHAPTER V

CONSTITUTION OF THE ATOM

We have seen that whether we produce the corpuscles by cathode rays, by ultra-violet light, or from incandescent metals, and whatever may be the metals or gases present we always get the same kind of corpuscles. Since corpuscles similar in all respects may be obtained from very different agents and materials, and since the mass of the corpuscles is less than that of any known atom, we see that the corpuscle must be a constituent of the atom of many different substances. That in fact the atoms of these substances have something in common.

We are thus confronted with the idea that the atoms of the chemical elements are built up of simpler systems; an idea which in various forms has been advanced by more than one chemist. Thus Prout, in 1815, put forward the view that the atoms of all the chemical elements are built up of atoms of hydrogen; if this were so the combining weights of all the elements would, on the assumption that there was no loss of weight when the atoms of hydrogen combined to form the atom of some other element, be integers; a result not in accordance with observation. To avoid this discrepancy Dumas suggested that the primordial atom might not be the hydrogen atom, but a smaller atom having only one-half or one-quarter of the mass of the hydrogen atom. Further support was given to the idea of the complex nature of the atom by the discovery by Newlands and Mendeleeff of what is known as the periodic law, which shows that there is a periodicity in the properties of the elements when they are arranged in the order of increasing atomic weights. The simple relations which exist between the combining weights of several of the elements having similar chemical properties, for example, the fact that the combining weight of sodium is the arithmetic mean of those of lithium and potassium, all point to the conclusion that the atoms of the different elements have something in common. Further evidence in the same direction is afforded by the similarity in the structure of the spectra of elements in the same group in the periodic series, a similarity which recent work on the existence in spectra of series of lines whose frequencies are connected by definite numerical relations has done much to emphasize and establish; indeed spectroscopic evidence alone has led Sir Norman Lockyer for a long time to advocate the view that the elements are really compounds which can be dissociated when the circumstances are suitable. The phenomenon of radio-activity, of which I shall have to speak later, carries the argument still further, for there seems good reasons for believing that radio-activity is due to changes going on within the atoms of the radio-active substances. If this is so then we must face the problem of the constitution of the atom, and see if we can imagine a model which has in it the potentiality of explaining the remarkable properties shown by radio-active substances. It may thus not be superfluous to consider the bearing of the existence of corpuscles on the problem of the constitution of the atom; and although the model of the atom to which we are led by these considerations is very crude and imperfect, it may perhaps be of service by suggesting lines of investigations likely to furnish us with further information about the constitution of the atom.

The Nature of the Unit from which the Atoms are Built Up

Starting from the hypothesis that the atom is an aggregation of a number of simpler systems, let us consider what is the nature of one of these systems. We have seen that the corpuscle, whose mass is so much less than that of the atom, is a constituent of the atom, it is natural to regard the corpuscle as a constituent of the primordial system. The corpuscle, however, carries a definite charge of negative electricity, and since with any charge of electricity we always associate an equal charge of the opposite kind, we should expect the negative charge on the corpuscle to be associated with an equal charge of positive electricity. Let us then take as our primordial system an electrical doublet, with a negative corpuscle at one end and an equal positive charge at the other, the two ends being connected by lines of electric force which we suppose to have a material existence. For reasons which will appear later on, we shall suppose that the volume over which the positive electricity is spread is very much larger than the volume of the corpuscle. The lines of force will therefore be very much more condensed near the corpuscle than at any other part of the system, and therefore the quantity of ether bound by the lines of force, the mass of which we regard as the mass of the system, will be very much greater near the corpuscle than elsewhere. If, as we have supposed, the size of the corpuscle is very small compared with the size of the volume occupied by the positive electrification, the mass of the system will practically arise from the mass of bound ether close to the corpuscle; thus the mass of the system will be practically independent of the position of its positive end, and will be very approximately the mass of the corpuscles if alone in the field. This mass (see page 21) is for each corpuscle equal to $\frac{2e^2}{3a}$, where e is the charge on the corpuscle and a its radius—a, as we have seen, being about 10^{-13} cm.

Now suppose we had a universe consisting of an immense number of these electrical doublets, which we regard as our primordial system; if these were at rest their mutual attraction would draw them together, just as the attractions of a lot of little magnets would draw them together if they were free to move, and aggregations of more than one system would be formed.

If, however, the individual systems were originally moving with considerable velocities, the relative velocity of two systems, when they came near enough to exercise appreciable attraction on each other, might be sufficient to carry the systems apart in spite of their mutual attraction. In this case the formation of aggregates would be postponed, until the kinetic energy of the units had fallen so low that when they came into collision, the tendency to separate due to their relative motion was not sufficient to prevent them remaining together under their mutual attraction.

Let us consider for a moment the way in which the kinetic energy of such an assemblage of units would diminish. We have seen (p. 68) that whenever the velocity of a charged body is changing the body is losing energy, since it generates electrical waves which radiate through space, carrying energy with them. Thus, whenever the units come into collision, *i.e.*, whenever they come so close together that they sensibly accelerate or retard each other's motion, energy will be radiated away, the whole of which will not be absorbed by the surrounding units. There will thus be a steady loss of kinetic energy, and after a time, although it may be a very long time, the kinetic energy will fall to the value at which aggregation of the units into groups of two will begin; these will later on be followed by the formation of aggregates containing a larger number of units.

In considering the question of the further aggregation of these complex groups, we must remember that the possibility of aggregation will depend not merely upon the velocity of the aggregate as a whole, *i.e.*, upon the velocity of the centre of gravity, but also upon the relative velocities of the corpuscles within the aggregate.

Let us picture to ourselves the aggregate as, like the Æpinus atom of Lord Kelvin, consisting of a sphere of uniform positive electrification, and exerting therefore a radial electric force proportional at an internal point to the distance from the centre, and that the very much smaller negatively electrified corpuscles are moving about inside it. The number of corpuscles is the number of units which had gone to make up the aggregate, and the total negative electrification on the corpuscles is equal to the positive electrification on the sphere. To fix our ideas let us take the case shown in Fig. 15 of three corpuscles

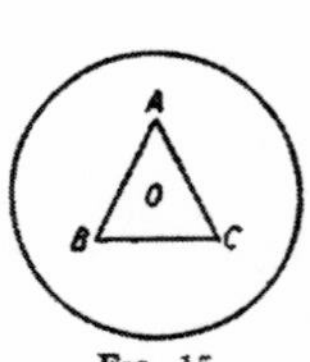

FIG. 15

A, *B*, *C*, arranged within the sphere at the corners of an equilateral triangle, the centre of the triangle coinciding with the centre of the sphere. First suppose the corpuscles are at rest; they will be in equilibrium when they are at such a distance from the centre of the sphere that the repulsion between the corpuscles, which will evidently be radial, just balances the radial attraction excited on the corpuscles by the positive electrification of the sphere. A simple calculation shows that this will be the case when the distance of the corpuscle from the centre is equal to .57 times the radius of the sphere. Next suppose that the corpuscles, instead of being at rest, are describing circular orbits round the centre of the sphere. Their centrifugal force will carry them farther away from the centre by an amount depending upon the speed with which they are rotating in their orbits. As we increase this speed the distance of the corpuscles from the centre of the sphere will increase until at a certain speed the corpuscles will reach the surface of the sphere; further increases in speed will cause them first to rotate outside the sphere and finally leave the sphere altogether, when the atom will break up.

In this way we see that the constitution of the

aggregate will not be permanent, if the kinetic energy due to the velocity of the corpuscles inside the sphere relative to the centre of the sphere exceeds a certain value. We shall, for the sake of brevity, speak of this kinetic energy of the corpuscles within the atom as the *corpuscular temperature* of the atom, and we may express the preceding result by saying that the atom will not be stable unless its corpuscular temperature is below a certain value.

We must be careful to distinguish between corpuscular temperature, which is the mean kinetic energy of the corpuscles inside the atom, and the molecular temperature, which is the mean kinetic energy due to the motion of the centre of gravity of the atom. These temperatures are probably not in any very close relationship with each other. They would be proportional to each other if the law known as the law of equipartition of energy among the various degrees of freedom of the atom were to apply. This law is, however, inconsistent with the physical properties of gases, and in the proof given of it in the kinetic theory of gases, no estimate is given of the time required to establish the state contemplated by the law; it may be that this time is so long that gases are never able to get into this state.

Let us now take the case of two aggregations, A and B, whose corpuscular temperatures are high, though not so high, of course, as to make A and B unstable when apart, and suppose, in order to give them the best possible chance of combining, that the centres of gravity of A and B when quite close to each other are at rest, will A and B unite to form a more complex aggregate as they would if the corpuscles in them were at rest? We can easily, I think, see that they will not necessarily do so. For as A and B approach each other, under their mutual attractions, the potential energy due to the separation of A and B will diminish and their kinetic energy will increase. This increase in the kinetic energy of the corpuscles in A and B will increase the tendency of the corpuscles to leave their atoms, and if the increase in the kinetic energy is considerable A and B may each lose one or more corpuscles. The departure of a corpuscle will leave A and B positively charged, and they will tend to separate under the repulsion of these charges. When separated they will have each a positive charge; but as there are now free corpuscles with negative charges moving about in the region in which A and B are situated, these positive charges will ultimately be neutralized by corpuscles striking against A and B and remaining in combination with them.

We thus conclude that unless the corpuscular temperature after union is less than a certain limiting value, the union cannot be permanent, the complex formed being unstable, and incapable of a permanent existence. Now, the corpuscular temperature of the aggregate formed by A and B will depend upon the corpuscular temperatures of A and B before union, and also upon the diminution in the potential energy of the system occasioned by the union of A and B. If the corpuscular temperatures of A and B before union were very high, the corpuscular temperature after union would be high also; if they were above a certain limit, the corpuscular temperature after union would be too high for stability, and the aggregate AB would not be formed. Thus, one condition for the formation of complex aggregates is that the corpuscular temperature of their constituents before combination should be sufficiently low.

If the *molecular* temperature of the gas in which A and B are molecules is very high, combination may be prevented by the high relative velocity of A and B carrying them apart in spite of their mutual attraction. The point, however, which I wish to emphasize is, that we cannot secure the union merely by lowering the molecular temperature, *i.e.*, by cooling the gas; union will be impossible unless the *corpuscular* temperature, *i.e.*, the kinetic energy due to the motion of the corpuscles inside the atom, is reduced below a certain value. We may prevent union by raising the molecular temperature of a gas, but we cannot ensure union by lowering it.

Thus, to take a specific example, the reason, on this view, why the atoms of hydrogen present on the earth do not combine to form some other element, even at the exceedingly low temperature at which hydrogen becomes liquid, is that even at this temperature the kinetic energy of the corpuscles inside the atom, *i.e.*, the corpuscular temperature, is too great. It may be useful to repeat here what we stated before, that there is no very intimate connection between the corpuscular and molecular temperatures, and that we may reduce the latter almost to the absolute zero without greatly affecting the former.

We shall now proceed to discuss the bearing of these results on the theory that the different chemical elements have been gradually evolved by the aggregation of primordial units.

Let us suppose that the first stage has been reached and that we have a number of systems formed by the union of two units. When first these binary systems, as we shall call them, were formed, the corpuscles in the system would have a considerable amount of kinetic energy. This would be so, because when the two units have come together there must be an amount of kinetic energy produced equal to the diminution in the potential energy consequent upon the coalescence of the two units. As these binary systems have originally high corpuscular temperatures they will not be likely to combine with each other or with another unit; before they can do so the kinetic energy of the corpuscles must get reduced.

We shall proceed immediately to discuss the way in which this reduction is effected, but we shall anticipate the result of the discussion by saying that it leads to the result that the rate of decay in the corpuscular temperature probably varies greatly from one binary system to another.

Some of the systems will therefore probably have reached a condition in which they are able to combine with each other or with a single unit long before others are able to do so. The systems of the first kind will combine, and thus we shall have systems formed, some of which contain three, others four units, while at the same time there are many of the binary systems left. Thus, the appearance of the more complex systems need not be simultaneous with the disappearance of all the simpler ones.

The same principle will apply to the formation of further aggregations by the systems containing three or four units; some of these will be ready to unite before the others, and we may have systems containing eight units formed before the more persistent of those containing four, three, two or even one unit have disappeared. With the further advance of aggregation the number of different systems present at one and the same time will increase.

Thus, if we regard the systems containing different numbers of units as corresponding to the different chemical elements, then as the universe gets older elements of higher and higher atomic weight may be expected to appear. Their appearance, however, will not involve the annihilation of the elements of lower atomic weight. The number of atoms of the latter will, of course, diminish, since the heavier elements are by hypothesis built up of material furnished by the lighter. The whole of the atoms of the latter would not, however, all be used up at once, and thus we may have a very large number of elements existing at one and the same time.

If, however, there is a continual fall in the corpuscular temperature of the atoms through radiation, the lighter elements will disappear in time, and unless there is disintegration of the heavier atoms, the atomic weight of the lightest element surviving will continually increase. On this view, since hydrogen is the lightest known element and the atom of hydrogen contains about a thousand corpuscles, all aggregations of less than a thousand units have entered into combination and are no longer free.

The way the Corpuscles in the Atom Lose or Gain Kinetic Energy

If the kinetic energy arising from the motion of the corpuscles relatively to the centre of gravity of the atom could by collisions be transformed into kinetic energy due to the motion of the atom as a whole, *i.e.*, into molecular temperature, it would follow from the kinetic theory of gases, since the number of corpuscles in the atom is exceedingly large, that the specific heat of a gas at constant pressure would be very nearly equal to the specific heat at constant volume; whereas, as a matter of fact, in no gas is there any approach to equality in these specific heats. We conclude, therefore, that it is not by collisions that the kinetic energy of the corpuscles is diminished.

We have seen, however (page 68), that a moving electrified particle radiates energy whenever its velocity is changing either in magnitude or direction. The corpuscles in the atom will thus emit electric waves, radiating energy and so losing kinetic energy.

The rate at which energy is lost in this way by the corpuscles varies very greatly with the number of the corpuscles and the way in which they are moving. Thus, if we have a single corpuscle describing a circular orbit of radius a with uniform velocity v, the loss of energy due to radiation per second is $\frac{2}{3}\frac{e^2v^4}{Va^2}$, where e is the charge on the corpuscles and V the velocity of light. If instead of a single corpuscle we had two corpuscles at opposite ends of a diameter moving round the same orbit with the same velocity as the single corpuscle, the loss of energy per second from the two would be very much less than from the single corpuscle, and the smaller the velocity of the corpuscle the greater would be the diminution in the loss of energy produced by increasing the number of corpuscles. The effect produced by increasing the number of corpuscles is shown in the following table, which gives the rate of radiation for each corpuscle for various numbers of corpuscles arranged at equal angular intervals round the circular orbit.

The table applies to two cases; in one the velocity of the corpuscles is taken as one-tenth that of light, and in the second as one-hundredth. The radiation from a single corpuscle is in each case taken as unity.

Number of corpuscles.	Radiation from each corpuscle.	
	$v = \frac{V}{10}$	$v = \frac{V}{100}$
1	1	1
2	9.6×10^{-2}	9.6×10^{-4}
3	4.6×10^{-3}	4.6×10^{-7}
4	1.7×10^{-4}	1.7×10^{-10}
5	5.6×10^{-5}	5.6×10^{-13}
6	1.6×10^{-7}	1.6×10^{-17}

Thus, we see that the radiation from each of a group of six corpuscles moving with one-tenth the velocity of light is less than one-five-millionth part of the radiation from a single corpuscle, describing the same orbit with the same velocity, while, when the velocity of the corpuscles is only one-hundredth of that of light, the reduction in the radiation is very much greater.

If the corpuscles are displaced from the symmetrical position in which they are situated at equal intervals round a circle whose centre is at rest, the rate of radiation will be very much increased. In the case of an atom containing a large number of corpuscles the variation in the rate at which energy is radiated will vary very rapidly with the way the corpuscles are moving about in the atom. Thus, for example, if we had a large number of corpuscles following close on one another's heels round a circular orbit the radiation would be exceedingly small; it would vanish altogether if the corpuscles were so close together that they formed a continuous ring of negative electrification. If the same number of particles were moving about irregularly in the atom, then though the kinetic energy possessed by the corpuscles in the second case might be no greater than in the first, the rate of radiation, *i.e.*, of corpuscular cooling, would be immensely greater.

Thus, we see that in the radiation of energy from corpuscles whose velocity is not uniform we have a process going on which will gradually cool the corpuscular temperature of the atom, and so, if the view we have been discussing is correct, enable the atom to form further aggregations and thus tend to the formation of new chemical elements.

This cooling process must be an exceedingly slow one, for although the corpuscular temperature when the atom of a new element is formed is likely to be exceedingly high, and the lowering in that temperature required before the atom can enter again into fresh aggregations very large, yet we have evidence that some of the elements must have existed unchanged for many thousands, nay, millions of years; we have, indeed, no direct evidence of any change at all in the atom. I think, however, that some of the phenomena of radio-activity to which I shall have to allude later, afford, I will not say a proof of, but a very strong presumption in favor of some such secular changes taking place in the atom.

We must remember, too, that the corpuscles in any atom are receiving and absorbing radiation from other atoms. This will tend to raise the corpuscular temperature of the atom and thus help to lengthen the time required for that temperature to fall to the point where fresh aggregations of the atom may be formed.

The fact that the rate of radiation depends so much upon the way the corpuscles are moving about in the atom indicates that the lives of the different atoms of any particular element will not be equal; some of these atoms will be ready to enter upon fresh changes long before the others. It is important to realize how large are the amounts of energy involved in the formation of a complex atom or in any rearrangement of the configuration of the corpuscles inside it. If we have an atom containing n corpuscles each with a charge e measured in electrostatic units, the total quantity of negative electricity in the atom is $n e$ and there is an equal quantity of positive electricity distributed through the sphere of positive electrification; hence, the work required to separate the atom into its constituent units will be comparable with $\frac{(n e)^2}{a}$, a being the radius of the sphere containing the corpuscles. Thus, as the atom has been formed by the aggregation of these units $\frac{(n e)^2}{a}$ will be of the same order of magnitude as the kinetic energy imparted to those constituents during their whole history, from the time they started as separate units, down to the time they became members of the atom under consideration. They will in this period have radiated away a large quantity of this energy, but the following calculation will show what an enormous amount of kinetic energy the corpuscles in the atom must possess even if they have only retained an exceedingly small fraction of that communicated to them. Let us calculate the value of $\frac{(ne)^2}{a}$ for all the atoms in a gram of the substance; let N be the number of these atoms in a gram, then $N\frac{(ne)^2}{a}$ is the value of the energy acquired by these atoms. If M is the mass of an atom $NM = 1$, thus:

$$N\frac{(ne)^2}{a} = \frac{1}{M}\frac{(ne)^2}{a};$$

but if m is the mass of a corpuscle

$$nm = M,$$

and therefore

$$N\frac{(ne)^2}{a} = \frac{e}{m}\frac{ne}{a};$$

now when e is measured in electrostatic units

$$\frac{e}{m} = 3 \times 10^{17} \text{ and } e = 3.4 \times 10^{-10};$$

and therefore

$$N \frac{(n e)^2}{a} = 10.2 \times 10^7 \times \frac{n}{a}. \qquad (1)$$

Let us take the case of the hydrogen atom for which $n = 1000$, and take for a the value usually assumed in the kinetic theory of gases for the radius of the atom, *i.e.*, 10^{-8} cm. then

$$N \frac{(n e)^2}{a} = 1.02 \times 10^{19} \text{ ergs};$$

this amount of energy would be sufficient to lift a million tons through a height considerably exceeding one hundred yards. We see, too, from (1) that this energy is proportional to the number of corpuscles, so that the greater the molecular weight of an element, the greater will be the amount of energy stored up in the atoms in each gram.

We shall return to the subject of the internal changes in the atom when we discuss some of the phenomena of radio-activity, but before doing so it is desirable to consider more closely the way the corpuscles arrange themselves in the atom. We shall begin with the case where the corpuscles are at rest. The corpuscles are supposed to be in a sphere of uniform positive electrification which produces a radial attractive force on each corpuscle proportional to its distance from the centre of the sphere, and the problem is to arrange the corpuscles in the sphere so that they are in equilibrium under this attraction and their mutual repulsions. If there are only two corpuscles, A B, we can see at once that they will be in equilibrium if placed so that A B and the centre of the sphere are in the same straight line and $OA = OB = \frac{1}{2}$ the radius of the sphere.

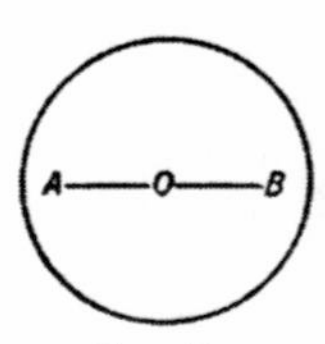

FIG. 16.

If there are three corpuscles, A B C, they will be in equilibrium of A B C as an equilateral triangle with its centre at O and $OA = OB = OC = (\frac{1}{5})^{\frac{1}{3}}$, or .57 times the radius of the sphere.

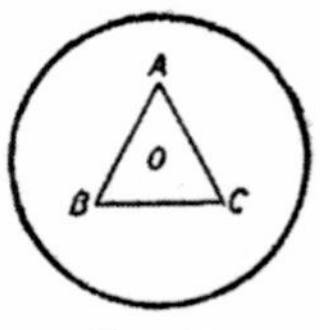

FIG. 15.

If there are four corpuscles these will be in equilibrium if placed at the angular points of a regular tetrahedron with its centre at the centre of the sphere. In these cases the corpuscles are all on the surface of a sphere concentric with the sphere of positive electrification, and we might suppose that whatever the number of corpuscles the position of equilibrium would be one of symmetrical distribution over the surface of a sphere. Such a distribution would indeed technically be one of equilibrium, but a mathematical calculation shows that unless the number of corpuscles is quite small, say seven or eight at the most, this arrangement is unstable and so can never persist. When the number of corpuscles is greater than this limiting number, the corpuscles break up into two groups. One group containing the smaller number of corpuscles is on the surface of a small body concentric with the sphere; the remainder are on the surface of a larger concentric body. When the number of corpuscles is still further increased there comes a stage when the equilibrium cannot be stable even with two groups, and the corpuscles now divide themselves into three groups, arranged on the surfaces of concentric shells; and as we go on increasing the number we pass through stages in which more and more groups are necessary for equilibrium. With any considerable number of corpuscles the problem of finding the distribution when in equilibrium becomes too complex for calculation; and we have to turn to experiment and see if we can make a model in which the forces producing equilibrium are similar to those we have supposed to be at work in the corpuscle. Such a model is afforded by a very simple and beautiful experiment first made, I think, by Professor Mayer. In this experiment a number of little magnets are floated in a vessel of water. The magnets are steel needles magnetized to equal strengths and are floated by being thrust through small disks of cork. The magnets are placed so that the positive poles are either all above or all below the surface of the water. These positive poles, like the corpuscles, repel each other with forces varying inversely as the distance between them. The attractive force is provided by a negative pole (if the little magnets have their positive poles above the water) suspended some distance above the surface of the water. This pole will exert on the positive poles of the little floating magnets an attractive force the component of which, parallel to the surface of the water, will be radial, directed to O, the projection of the negative pole on the surface of the water, and if the negative pole is some distance above the surface the component of the force to O will be very approximately proportional to the distance from O. Thus the forces on the poles of the floating magnets will be very similar to those acting on the corpuscle in our hypothetical atom;

the chief difference being that the corpuscles are free to move about in all directions in space, while the poles of the floating magnets are constrained to move in a plane parallel to the surface of the water.

The configurations which the floating magnets assume as the number of magnets increases from two up to nineteen is shown in Fig. 17, which was given by Mayer.

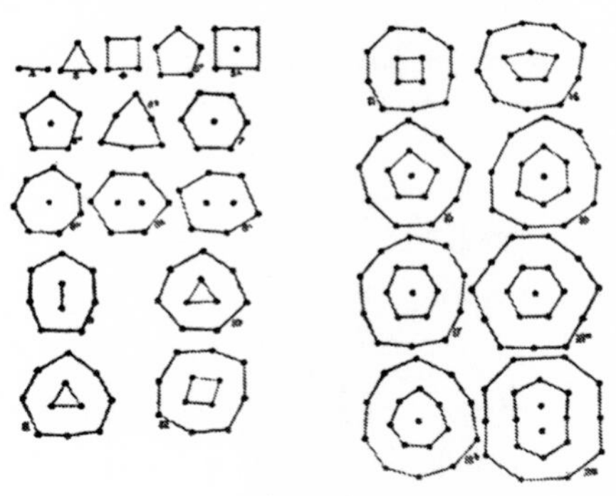

FIG. 17.

The configuration taken up when the magnets are more numerous can be found from the following table, which is also due to Mayer. From this table it will be seen that when the number of floating magnets does not exceed five the magnets arrange themselves at the corners of a regular polygon, five at the corners of a pentagon, four at the corners of a square and so on. When the number is greater than five this arrangement no longer holds. Thus, six magnets do not arrange themselves at the corners of a hexagon, but divide into two systems, one magnet being at the centre and five outside it at the corners of a regular pentagon. This arrangement in two groups lasts until there are fifteen magnets, when we have three groups; with twenty-seven magnets we get four groups and so on.

Arrangement of Magnets (Mayer)

1.	2.	3.	4.	5.
1 . 5	2 . 6	3 . 7	4 . 8	5 . 9
1 . 6	2 . 7	3 . 8	4 . 9	
1 . 7				
1 . 5 . 9	2 . 7 . 10	3 . 7 . 10	4 . 8 . 12	5 . 9 . 12
1 . 6 . 9	2 . 8 . 10	3 . 7 . 11	4 . 8 . 13	5 . 9 . 13
1 . 6 . 10	2 . 7 . 11	3 . 8 . 10	4 . 9 . 12	
1 . 6 . 11		3 . 8 . 11	4 . 9 . 13	
		3 . 8 . 12		
		3 . 8 . 13		

1.	2.	3.	4.
1 . 5 . 9 . 12	2 . 7 . 10 . 15	3 . 7 . 12 . 13	4 . 9 . 13 . 14
1 . 5 . 9 . 13	2 . 7 . 12 . 14	3 . 7 . 12 . 14	4 . 9 . 13 . 15
1 . 6 . 9 . 12		3 . 7 . 13 . 14	4 . 9 . 14 . 15
1 . 6 . 10 . 12		3 . 7 13 . 15	
1 . 6 . 10 . 13			
1 . 6 . 11 . 2			
1 . 6 . 11 . 13			
1 . 6 . 11 . 14			
1 . 6 . 11 . 15			

Where, for example, 3. 7. 12. 13 means that thirty-five magnets arrange themselves so that there is a ring of three magnets inside, then a ring of seven, then one of twelve, and one of thirteen outside.

I think this table affords many suggestions toward the explanation of some of the properties possessed by atoms. Let us take, for example, the chemical law called the Periodic Law; according to this law if we arrange the elements in order of increasing atomic weights, then taking an element of low atomic weight, say lithium, we find certain properties associated with it. These properties are not possessed by the elements immediately following it in the series of increasing atomic weight; but they appear again when we come to sodium, then they disappear again for a time, but reappear when we reach potassium, and so on. Let us now consider the arrangements of the floating magnets, and suppose that the number of magnets is proportional to the combining weight of an element. Then, if any property were associated with the triangular arrangement of magnets, it would be possessed by the elements whose combining weight was on this scale three, but would not appear again until we reached the combining weight ten, when it reappears, as for ten magnets we have the triangular arrangement in the middle and a ring of seven magnets outside. When the number of magnets is increased the triangular arrangement disappears for a time, but reappears with twenty magnets, and again with thirty-five, the triangular arrangement appearing and disappearing in a way analogous to the behavior of the properties of the elements in the Periodic Law. As an example of a property that might very well be associated with a particular grouping of the corpuscles, let us take the times of vibration of the system, as shown by the position of the lines in the spectrum of the element. First let us take the case of three corpuscles by themselves in the positively electrified sphere. The three corpuscles have nine degrees of freedom, so that there are nine possible periods. Some of these periods in this case would be infinitely long, and several of the possible periods would be equal to each other, so that we should not get nine different periods.

Suppose that the lines in the spectrum of the three corpuscles are as represented in Fig. 18 *a*,

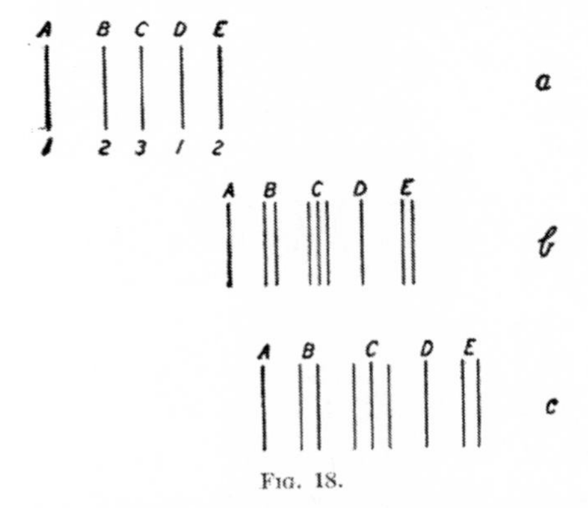

Fig. 18.

where the figures under the lines represent the number of periods which coalesce at that line; *i.e.*, regarding the periods as given by an equation with nine roots, we suppose that there is only one root giving the period corresponding to the line A, while corresponding to B there are two equal roots, three equal roots corresponding to C, one root, to *O*, and two to *E*. These periods would have certain numerical relations to each other, independent of the charge on the corpuscle, the size of the sphere in which they are placed, or their distance from the centre of the sphere. Each of these quantities, although it does not affect the ratio of the periods, will have a great effect upon the absolute value of any one of them. Now, suppose that these three corpuscles, instead of being alone in the sphere, form but one out of several groups in it, just as the triangle of magnets forms a constituent of the grouping of 3, 10, 20, and 35 magnets. Let us consider how the presence of the other groups would affect the periods of vibration of the three corpuscles. The absolute values of the periods would generally be entirely different, but the relationship existing between the various periods would be much more persistent, and although it might be modified it would not be destroyed. Using the phraseology of the Planetary Theory, we may regard the motion of the three corpuscles as "disturbed" by the other groups.

When the group of three corpuscles was by itself there were several displacements which gave the same period of vibration; for example, corresponding to the line *C* there were three displacements, all giving the same period. When, however, there are other groups present, then these different displacements will no longer be symmetrical with respect to these groups, so that the three periods will no longer be quite equal. They would, however, be very nearly equal unless the effect of the other groups is very large. Thus, in the spectrum, *C*, instead of being a single line, would become a triplet, while *B* and *E* would become doublets. *A D* would remain single lines.

Thus, the spectrum would now resemble Fig. 18 *b*; the more groups there are surrounding the group of three the more will the motion of the latter be disturbed and the greater the separation of the constituents of the triplets and doublets. The appearance as the number of groups increases is shown in Fig. 18 *b*, *c*. Thus, if we regarded the element which contain this particular grouping of corpuscles as being in the same group in the classification of elements according to the Periodic Law, we should get in the spectra of these elements homologous series of lines, the distances between the components of the doublets and triplets increasing with the atomic weight of the elements. The investigations of Rydberg, Runge and Paschen and Keyser have shown the existence in the spectra of elements of the same group series of lines having properties in many respects analogous to those we have described.

Another point of interest given by Mayer's experiments is that there is more than one stable configuration for the same number of magnets; these configurations correspond to different amounts of potential energy, so that the passage from the configuration of greater potential energy to that of less would give kinetic energy to the corpuscle. From the values of the potential energy stored in the atom, of which we gave an estimate on page 111, we infer that a change by even a small fraction in that potential energy would develop an amount of kinetic energy which if converted into heat would greatly transcend the amount of heat developed when the atoms undergo any known chemical combination.

An inspection of the table shows that there are certain places in it where the nature of the configuration changes very rapidly with the number of magnets; thus, five magnets form one group, while six magnets form two; fourteen magnets form two groups, fifteen three; twenty-seven magnets form three groups, twenty-eight four, and so on. If we arrange the chemical elements in the order of their atomic weights we find there are certain places where the difference in properties of consecutive elements is exceptionally great; thus, for example, we have extreme differences in properties between fluorine and sodium. Then there is more or less continuity in the properties until we get to chlorine, which is followed by potassium; the next break occurs at bromine and rubidium and so on. This effect seems analogous to that due to the regrouping of the magnets.

So far we have supposed the corpuscles to be at rest; if, however, they are in a state of steady motion and describing circular orbits round the centre of the sphere, the effect of the centrifugal force arising from this motion will be to drive the corpuscles farther away from the centre of the sphere, without, in many cases, destroying the character of the configuration. Thus, for example, if we have three corpuscles in the sphere, they will, in the state of steady motion, as when they are at rest, be situated at the corners of an equiangular triangle; this triangle will, however, be rotating round the centre of the sphere, and the distance of the corpuscles from the centre will be greater than when they are at rest and will increase with the velocity of the corpuscles.

There are, however, many cases in which rotation is essential for the stability of the configuration. Thus, take the case of four corpuscles. These, if rotating rapidly, are in stable steady motion when at the corners of a square, the plane of the square being at right angles to the axis of rotation; when, however, the velocity of rotation of the corpuscles falls below a certain value, the arrangement of four corpuscles in one plane becomes unstable, and the corpuscles tend to place themselves at the corners of a regular tetrahedron, which is the stable arrangement when the corpuscles are at rest. The system of four corpuscles at the corners of a square may be compared with a spinning top, the top like the corpuscles being unstable unless its velocity of rotation exceeds a certain critical value. Let us suppose that initially the velocity of the corpuscles exceeds this value, but that in some way or another the corpuscles gradually lose their kinetic energy; the square arrangement will persist until the velocity of the corpuscles is reduced to the critical value. The arrangement will then become unstable, and there will be a convulsion in the system accompanied by a great evolution of kinetic energy.

Similar considerations will apply to many assemblages of corpuscles. In such cases the configuration when the corpuscles are rotating with great rapidity will (as in the case of the four corpuscles) be essentially different from the configuration of the same number of corpuscles when at rest. Hence there must be some critical velocity of the corpuscles, such that, for velocities greater than the critical one, a configuration is stable, which becomes unstable when the velocity is reduced below the critical value. When the velocity sinks below the critical value, instability sets in, and there is a kind of convulsion or explosion, accompanied by a great diminution in the potential energy and a corresponding increase in the kinetic energy of the corpuscles. This increase in the kinetic energy of the corpuscles may be sufficient to detach considerable numbers of them from the original assemblage.

These considerations have a very direct bearing on the view of the constitution of the atoms which we have taken in this chapter, for they show that with atoms of a special kind, *i.e.*, with special atomic weights, the corpuscular cooling caused by the radiation from the moving corpuscles which we have supposed to be slowly going on, might, when it reached a certain stage, produce instability inside the atom, and produce such an increase in the kinetic energy of the corpuscles as to give rise to greatly increased radiation, and it might be detachment of a portion of the atom. It would cause the atom to emit energy; this energy being derived from the potential energy due to the arrangement of the corpuscles in the atom. We shall see when we consider the phenomenon of radio-activity that there is a class of bodies which show phenomena analogous to those just described.

On the view that the lighter elements are formed first by the aggregation of the unit doublet, the negative element of which is the corpuscle, and that it is by the combination of the atoms of the lighter elements that the atoms of the heavier elements are produced, we should expect the corpuscles in the heavy atoms to be arranged as it were in bundles, the arrangement of the corpuscles in each bundle being similar to the arrangement in the atom of some lighter element. In the heavier atom these bundles would act as subsidiary units, each bundle corresponding to one of the magnets in the model formed by the floating magnets, while inside the bundle themselves the corpuscle would be the analogue of the magnet.

We must now go on to see whether an atom built up in the way we have supposed could possess any of the properties of the real atom. Is there, for example, in this model of an atom any scope for the electro-chemical properties of the real atom; such properties, for example, as those illustrated by the division of the chemical elements into two classes, electro-positive and electro-negative. Why, for example, if this is the constitution of the atom, does an atom of sodium or potassium tend to acquire a positive, the atom of chlorine a negative charge of electricity? Again, is there anything in the model of the atom to suggest the possession of such a property as that called by the chemists valency; *i.e.*, the property which enables us to divide the elements into groups, called monads, dyads, triads, such that in a compound formed by any two elements of the first group the molecule of the compound will contain the same number of atoms of each element, while in a compound formed by an element A in the first group with one B in the second, the molecule of the compound contains twice as many atoms of *A* as of *B*, and so on?

Let us now turn to the properties of the model atom. It contains a very large number of corpuscles in rapid motion. We have evidence from the phenomena connected with the conduction of electricity through gases that one or more of these corpuscles can be detached from the atom. These may escape owing to their high velocity enabling them to travel beyond the attraction of the atom. They may be detached also by collision of the atom with other rapidly moving atoms or free corpuscles. When once a corpuscle has escaped from an atom the latter will have a positive charge. This will make it more difficult for a second negatively electrified corpuscle to escape, for in consequence of the positive charge on the atom the latter will attract the second corpuscle more strongly than it did the first. Now we can readily conceive that the ease with which a particle will escape from, or be knocked out of, an atom may vary very much in the atoms of the different elements. In some atoms the velocities of the corpuscles may be so great that a corpuscle escapes at once from the atom. It may even be that after one has escaped, the attraction of the positive electrification thus left on the atom is not sufficient to restrain a second, or even a third, corpuscle from escaping. Such atoms would acquire positive charges of one, two, or three units, according as they lost one, two, or three corpuscles. On the other hand, there may be atoms in which the velocities of the corpuscles are so small that few, if any, corpuscles escape of their own accord, nay, they may even be able to receive one or even more than one corpuscle before the repulsion exerted by the negative electrification on these foreign corpuscles forces any of the original corpuscles out. Atoms of this kind if placed in a region where corpuscles were present would by aggregation with these corpuscles receive a negative charge. The magnitude of the negative charge would depend upon the firmness with which the atom held its corpuscles. If a negative charge of one corpuscle were not sufficient to expel a corpuscle while the negative charge of two corpuscles could do so, the maximum negative charge on the atom would be one unit. If two corpuscles were not sufficient to expel a corpuscle, but three were, the maximum negative charge would be two units, and so on. Thus, the atoms of this class tend to get charged with negative electricity and correspond to the electro-negative chemical elements, while the atoms of the class we first considered, and which readily lose corpuscles, acquire a positive charge and correspond to the atoms of the electro-positive elements. We might conceive atoms in which the equilibrium of the corpuscles was so nicely balanced that though they do not of themselves lose a corpuscle, and so do not acquire a positive charge, the repulsion exerted by a foreign corpuscle coming on to the atom would be sufficient to drive out a corpuscle. Such an atom would be incapable of receiving a charge either of positive or negative electricity.

Suppose we have a number of the atoms that readily lose their corpuscles mixed with a number of those that can retain a foreign corpuscle. Let us call an atom of the first class A, one of the second B, and suppose that the A atoms are of the kind that lose one corpuscle while the B atoms are of the kind that can retain one, but not more than one; then the corpuscles which escape from the A atoms will ultimately find a home on the B atoms, and if there are an equal number of the two kinds of atoms present we shall get ultimately all the A atoms with the unit positive charge, all the B atoms with the unit negative charge. These oppositely electrified atoms will attract each other, and we shall get the compound A B formed. If the A atoms had been of the kind that lost two corpuscles, and the B atoms the same as before, then the A atoms would get the charge of two positive units, the B atoms a charge of one unit of negative electricity. Thus, to form a neutral system two of the B atoms must combine with one of the A's and thus the compound A B_2 would be formed.

Thus, from this point of view a univalent electro-positive atom is one which, under the circumstances prevailing when combination is taking place, has to lose one and only one corpuscle before stability is attained; a univalent electro-negative atom is one which can receive one but not more than one corpuscle without driving off other corpuscles from the atom; a divalent electro-positive atom is one that loses two corpuscles and no more, and so on. The valency of the atom thus depends upon the ease with which corpuscles can escape from or be received by the atom; this may be influenced by the circumstances existing when combination is taking place. Thus, it would be easier for a corpuscle, when once it had got outside the atom, to escape being pulled back again into it by the attraction of its positive electrification, if the atom were surrounded by good conductors than if it were isolated in space. We can understand, then, why the valency of an atom may in some degree be influenced by the physical conditions under which combination is taking place.

On the view that the attraction between the atoms in a chemical compound is electrical in its origin, the ability of an element to enter into chemical combination depends upon its atom having the power of acquiring a charge of electricity. This, on the preceding view, implies either that the uncharged atom is unstable and has to lose one or more corpuscles before it can get into a steady state, or else that it is so stable that it can retain one or more additional corpuscles without any of the original corpuscles being driven out. If the range of stability is such that the atom, though stable when uncharged, becomes unstable when it receives an additional corpuscle, the atom will not be able to receive a charge either of positive or negative electricity, and will therefore not be able to enter into chemical combination. Such an atom would have the properties of the atoms of such elements as argon or helium.

The view that the forces which bind together the atoms in the molecules of chemical compounds are electrical in their origin, was first proposed by Berzelius; it was also the view of Davy and of Faraday. Helmholtz, too, declared that the mightiest of the chemical forces are electrical in their origin. Chemists in general seem, however, to have made but little use of this idea, having apparently found the conception of "bonds of affinity" more fruitful. This doctrine of bonds is, however, when regarded in one aspect almost identical with the electrical theory. The theory of bonds when represented graphically supposes that from each univalent atom a straight line (the symbol of a bond) proceeds; a divalent atom is at the end of two such lines, a trivalent atom at the end of three, and so on; and that when the chemical compound is represented by a graphic formula in this way, each atom must be at the end of the proper number of the lines which represent the bonds. Now, on the electrical view of chemical combination, a univalent atom has one unit charge, if we take as our unit of charge the charge on the corpuscle; the atom is therefore the beginning or end of one unit Faraday tube: the beginning if the charge on the atom is positive, the end if the charge is negative. A divalent atom has two units of charge and therefore it is the origin or termination of two unit Faraday tubes. Thus, if we interpret the "bond" of the chemist as indicating a unit Faraday tube, connecting charged atoms in the molecule, the structural formulæ of the chemist can be at once translated into the electrical theory. There is, however, one point of difference which deserves a little consideration: the symbol indicating a bond on the chemical theory is not regarded as having direction; no difference is made on this theory between one end of a bond and the other. On the electrical theory, however, there is a difference between the ends, as one end corresponds to a positive, the other to a negative charge. An example or two may perhaps be the easiest way of indicating the effect of this consideration. Let us take the gas ethane whose structural formula is written

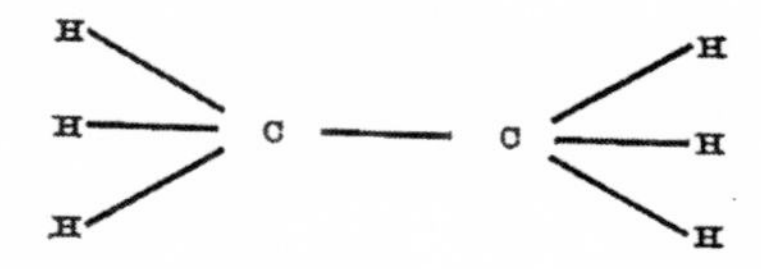

According to the chemical view there is no difference between the two carbon atoms in this compound; there would, however, be a difference on the electrical view. For let us suppose that the hydrogen atoms are all negatively electrified; the three Faraday tubes going from the hydrogen atoms to each carbon atom give a positive charge of three units on each carbon atom. But in addition to the Faraday tubes coming from the hydrogen atoms, there is one tube which goes from one carbon atom to the other. This means an additional positive charge on one carbon atom and a negative charge on the other. Thus, one of the carbon atoms will have a charge of four positive units, while the other will have a charge of three positive and one negative unit, *i.e.*, two positive units; so that on this view the two carbon atoms are not in the same state. A still greater difference must exist between the atoms when we have what is called double linking, *i.e.*, when the carbon atoms are supposed to be connected by two bonds, as in the compound

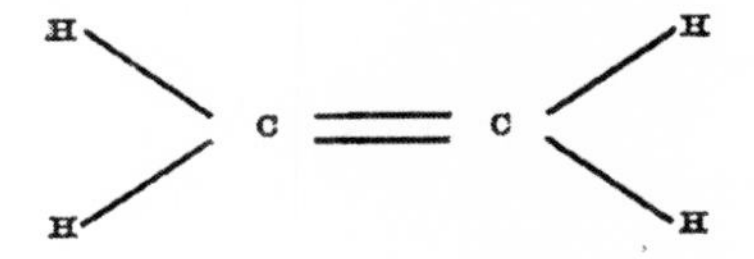

Here, if one carbon atom had a charge of four positive units, the other would have a charge of two positive and two negative units.

We might expect to discover such differences as are indicated by these considerations by the investigation of which are known as additive properties, *i.e.*, properties which can be calculated when the chemical constitution of the molecule is known. Thus, let A B C represent the atoms of three chemical elements, then if p is the value of some physical constant for the molecule of A_2, q the value for B_2, and r for C_2, then if this constant obeys the additive law, its value for a molecule of the substance whose chemical composition is represented by the formula A_x B_y C_z is

$$\tfrac{1}{2}px + \tfrac{1}{2}qy + \tfrac{1}{2}rz.$$

We can only expect relations like this to hold when the atoms which occur in the different compounds corresponding to different values of x y z are the same. If the atom A occurs in different states in different compounds we should have to use different values of p for these compounds.

A well-known instance of the additive property is the refractive power of different substances for light, and in this case chemists find it necessary to use different values for the refraction due a carbon atom according as the atom is doubly or singly linked. They use, however, the same value for the refraction of the carbon atom when singly linked with another atom as when, as in the compound CH_4, it is not linked with another carbon atom at all.

It may be urged that although we can conceive that one atom in a compound should be positively and the other negatively electrified when the atoms are of different kinds, it is not easy to do so when the atoms are of the same kind, as they are in the molecules of the elementary gases H_2, O_2, N_2 and so on. With reference to this point we may remark that the electrical state of an atom, depending as it does on the power of the atom to emit or retain corpuscles, may be very largely influenced by circumstances external to the atom. Thus, for an example, an atom in a gas when surrounded by rapidly moving atoms or corpuscles which keep striking against it may have corpuscles driven out of it by these collisions and thus become positively electrified. On the other hand, we should expect that, *ceteris paribus*, the atom would be less likely to lose a corpuscle when it is in a gas than when in a solid or a liquid. For when in a gas after a corpuscle has just left the atom it has nothing beyond its own velocity to rely upon to escape from the attraction of the positively electrified atom, since the other atoms are too far away to exert any forces upon it. When, however, the atom is in a liquid or a solid, the attractions of the other atoms which crowd round this atom may, when once a corpuscle has left its atom, help it to avoid falling back again into atom. As an instance of this effect we may take the case of mercury in the liquid and gaseous states. In the liquid state mercury is a good conductor of electricity. One way of regarding this electrical conductivity is to suppose that corpuscles leave the atoms of the mercury and wander about through the interstices between the atoms. These charged corpuscles when acted upon by an electric force are set in motion and constitute an electric current, the conductivity of the liquid mercury indicating the presence of a large number of corpuscles. When, however, mercury is in the gaseous state, its electrical conductivity has been shown by Strutt to be an exceedingly small fraction of the conductivity possessed by the same number of molecules when gaseous. We have thus indications that the atoms even of an electro-positive substance like mercury may only lose comparatively few corpuscles when in the gaseous state. Suppose then that we had a great number of atoms all of one kind in the gaseous state and thus moving about and coming into collision with each other; the more rapidly moving ones, since they would make the most violent collisions, would be more likely to lose corpuscles than the slower ones. The faster ones would thus by the loss of their corpuscles become positively electrified, while the corpuscles driven off would, if the atoms were not too electro-positive to be able to retain a negative charge even when in the gaseous state, tend to find a home on the more slowly moving atoms. Thus, some of the atoms would get positively, others negatively electrified, and those with changes of opposite signs would combine to form a diatomic molecule. This argument would not apply to very electro-positive gases. These we should not expect to form molecules, but since there would be many free corpuscles in the gas we should expect them to possess considerable electrical conductivity.

CHAPTER VI

RADIO-ACTIVITY AND RADIO-ACTIVE SUBSTANCES

In 1896 Becquerel discovered that uranium and its salts possess the power of giving out rays which, like Röntgen and cathode rays, affect a photographic plate, and make a gas through which they pass a conductor of electricity. In 1898 Schmidt discovered that thorium possesses similar properties. This power of emitting rays is called radio-activity, and substances which possess the power are said to be radio-active.

This property of uranium led to a careful examination of a large number of minerals containing this substance, and M. and Mme. Curie found that some of these, and notably some specimens of pitch-blende, were more radio-active than equal volumes of pure uranium, although only a fraction of these minerals consisted of uranium. This indicated that these minerals contained a substance or substances much more radio-active than uranium itself, and a systematic attempt was made to isolate these substances. After a long investigation, conducted with marvellous skill and perseverance, M. and Mme. Curie, with the collaboration of MM. Bemont and Debierne, succeeded in establishing the existence of three new radio-active substances in pitch-blende: radium associated with the barium in the mineral, and closely resembling it in its chemical properties; polonium associated with the bismuth, and actinium with the thorium. They succeeded in isolating the first of these and determined its combining weight, which was found to be 225. Its spectrum has been discovered and examined by Demarçay. Neither polonium nor actinium has yet been isolated, nor have their spectra been observed. The activity of polonium has been found to be fugitive, dying away in some months after its preparation.

These radio-active substances are not confined to rare minerals. I have lately found that many specimens of water from deep wells contain a radio-active gas, and Elster and Geitel have found that a similar gas is contained in the soil.

These radio-active substances may be expected to be of the greatest possible assistance in the task of investigating problems dealing with the nature of the atom, and with the changes that go on in the atom from time to time. For the properties possessed by these substances are so marked as to make the detection of exceedingly minute quantities of them a matter of comparative ease. The quantity of these substances which can be detected is to the corresponding amount of the other elements which have to be detected by the ordinary methods of chemical analysis, in the proportion of a second to thousands of years. Thus, changes which would have to go on for almost geological epochs with the non-radio-active substances, before they became large enough to be detected, could with radio-active substances prove appreciable effects in the course of a few hours.

Character of the Radiation

Rutherford found that the radiation from uran ium, and it has subsequently been found that the same is true for thorium and radium, is made up of three distinct types which he calls the α, β, and γ radiations.

The α radiation is very easily absorbed, being unable to penetrate more than a few millimetres of air at atmospheric pressure, the β radiation is much more penetrating, while the γ radiation is the most penetrating of all. Investigations of the effects of magnetic and electric forces on these three types of radiation have shown that they are of entirely different characters. Becquerel showed that the β rays were deflected by electric and magnetic forces, the direction of the deflection showing that the rays carried a charge of negative electricity. He determined, using the method described in Chapter IV, the value of $\frac{e}{m}$, the ratio of the charge to the mass of the carriers of the negative electricity; he found that it was about 10^7, and that the velocity for some of the rays was more than two-thirds that of light. He thus proved that the β rays consisted of corpuscles travelling at prodigious speeds.

The α rays are not nearly so easily deflected as the β rays, but Rutherford has recently shown that they can be deflected, and the direction of deflection shows that they carry a *positive* charge. He finds, and his measurements have been confirmed by Des Coudres, that the ratio of $\frac{e}{m}$ is 6×10^3, and the velocity of these particles is 2×10^9 centimetres per second. The value of $\frac{e}{m}$ shows that the carriers of the positive electrification have masses comparable with those of ordinary atoms; thus $\frac{e}{m}$ for hydrogen is 10^4 and for helium 2.5×10^3. The very high velocity with which these are shot out involves an enormous expenditure of energy, a point to which we shall return later. One of the most interesting things about this result is that the value of $\frac{e}{m}$ shows that the atoms shot off are not the atoms of radium, indicating either that radium is a compound containing lighter elements or else that the atom of radium is disintegrating into such elements. The value of $\frac{e}{m}$ for the α rays obtained by Rutherford and Des Coudres suggests the existence of a gas heavier than hydrogen but lighter than helium. The γ rays, as far as we know, are not deflected either by magnetic or electric forces.

There is considerable resemblance between a radio-active substance and a substance emitting secondary radiation under the influence of Röntgen rays: the secondary radiation is known to contain radiation of the β and γ types; and as part of the radiation is exceedingly easily absorbed, being unable to penetrate more than a millimetre or so of air at atmospheric pressure, it is possible that closer investigation may show that α rays, *i.e.*, positively electrified particles, are present also. This analogy raises the question as to whether there may not, in the case of the body struck by the Röntgen rays, be a liberation of energy such as we shall see occurs in the case of the radio-active substances, the energy emitted by the radiating substances being greater than the energy in the Röntgen rays falling upon it; this excess of energy being derived from changes taking place in the atoms of the body exposed to the Röntgen rays. This point seems worthy of investigation, for it might lead to a way of doing by external agency what radio-active bodies can do spontaneously, *i.e.*, liberate the energy locked up in the atom.

Emanation from Radio-Active Substances

Rutherford proved that thorium emits something which is radio-active and which is wafted about by currents of air as if it were a gas; in order to avoid prejudging the question as to the physical state in which the substance given off by radium exists, Rutherford called it the "emanation." The emanation can pass through water or the strongest acid and can be raised to temperatures at which platinum is incandescent without suffering any loss of radio-activity. In this inertness it resembles the gases argon and helium, the latter of which is almost always found associated with thorium. The radio-activity of the thorium emanation is very transient, sinking to half its value in about one minute.

The Curies found that radium also gives off a radio-active emanation which is much more persistent than that given off by thorium, taking about four days to sink to half its activity.

There seems every reason for thinking that those emanations are radio-active matter in the gaseous form; they can be wafted from one place to another by currents of air; like a gas they diffuse through a porous plug at a rate which shows that their density is very high. They diffuse gradually through air and other gases. The coefficient of diffusion of the radium emanation through air has been measured by Rutherford and Miss Brooks and they concluded that the density of the emanation was about eighty. The emanation of radium has been liquefied by Rutherford and Soddy; and I have, by the kindness of Professor Dewar, been able to liquefy the radio-active gas found in water from deep wells, which very closely resembles the emanation and is quite possibly identical with it. In short the emanations seem to satisfy every test of the gaseous state that can be applied to them. It is true that they are not capable of detection by any chemical tests of the ordinary type, nor can they be detected by spectrum analysis, but this is only because they are present in very minute quantities —quantities far too small to be detected even by spectrum analysis, a method of detection which is exceedingly rough when compared with the electrical methods which we are able to employ for radio-active substances. It is not, I think, an exaggeration to say that it is possible to detect with certainty by the electrical method a quantity of a radio-active substance less than one-hundred-thousandth part of the least quantity which could be detected by spectrum analysis.

Each portion of a salt of radium or thorium is giving off the emanation, whether that portion be on the inside or the outside of the salt; the emanation coming from the interior of a salt, however, does not escape into the air, but gets entangled in the salt and accumulates. If such a radio-active salt is dissolved in water, there is at first a great evolution of the emanation which has been stored up in the solid salt. The emanation can be extracted from the water either by boiling the water or bubbling air through it. The stored up emanation can also be driven off from salts in the solid state by raising them to a very high temperature.

Induced Radio-Activity

Rutherford discovered that substances exposed to the emanation from thorium become radio-active, and the Curies discovered almost simultaneously that the same property is possessed by the emanation from radium. This phenomenon is called induced radio-activity. The amount of induced radio-activity does not depend upon the nature of the substance on which it is induced; thus, paper becomes as radio-active as metal when placed in contact with the emanations of thorium or radium.

The induced radio-activity is especially developed on substances which are negatively electrified. Thus, if the emanation is contained in a closed vessel, in which a negatively electrified wire is placed, the induced radio-activity is concentrated on the negatively electrified wire, and this induced activity can be detected on negatively electrified bodies when it is too weak to be detected on unelectrified surfaces. The fact that the nature of the induced radio-activity does not depend on the substance in which it is induced points to its being due to a radio-active substance which is deposited from the emanation on substances with which it comes in contact.

Further evidence of this is afforded by an experiment made by Miss Gates, in which the induced radio-activity on a fine wire was, by raising it to incandescence, driven off the wire and deposited on the surrounding surfaces. The induced radio-activity due to the thorium emanation is very different from that due to the radium emanation, for whereas the activity of the thorium emanation is so transient that it drops to half its value in one minute, the induced radio-activity due to it takes about eleven hours to fall in the same proportion. The emanation due to radium, which is much more lasting than the thorium emanation, taking about four days instead of one minute to fall to half its value, gives rise to a very much less durable induced radio-activity, one falling to half its value in about forty minutes instead of, as in the case of thorium, eleven hours. The emanation due to actinium is said only to be active for a few seconds, but the induced radio-activity due to it seems to be nearly as permanent as that due to radium.

Separation of the Active Constituent from Thorium

Rutherford and Soddy, in a most interesting and important investigation, have shown that the radio-activity of thorium is due to the passage of the thorium into a form which they call *T h X*, which they showed could be separated from the rest of the thorium by chemical means. When this separation has been effected the thorium left behind is for a time deprived of most of its radio-activity, which is now to be found in the *T h X*. The radio-activity of the thorium X slowly decays while that of the rest of the thorium increases until it has recovered its original activity. While this has been going on, the radio-activity of the *T h X* has vanished. The time taken for the radio-activity of the *T h X* to die away to half its original value has been shown by Rutherford and Soddy to be equal to the time taken by the thorium from which the *T h X* has been separated to recover half its original activity. All these results support the view that the radio-active part of the thorium, the thorium X, is continually being produced from the thorium itself; so that if the activity of thorium X were permanent, the radio-activity of the thorium would continually increase. The radio-activity of the thorium X, however, steadily dies away. This prevents the unlimited increase of the radio-activity of the mixture, which will reach a steady value when the increase in the radio-activity due to the production of fresh *Th X* is balanced by the decay in the activity of that already produced. The question arises as to what becomes of the *Th X* and the emanation when they have lost their radio-activity. This dead *Th X*, as we may call it, is accumulating all the time in the thorium; but inasmuch as it has lost its radio-activity, we have only the ordinary methods of chemical analysis to rely upon, and as these are almost infinitely less delicate than the tests we can apply to radio-active substances, it might take almost geological epochs to accumulate enough of the dead *Th X* to make detection possible by chemical analysis. It seems possible that a careful examination of the minerals in which thorium and radium occur might yield important information. It is remarkable that helium is almost invariably a constituent of these minerals.

You will have noticed how closely, as pointed

out by Rutherford and Soddy, the production of radio-activity seems connected with changes taking place in the radio-active substance. Thus, to take the case of thorium, which is the one on which we have the fullest information, we have first the change of thorium into thorium X, then the change of the thorium X into the emanation and the substance forming the α rays. The radio-activity of the emanation is accompanied by a further transformation, one of the products being the substance which produces induced radio-activity.

On this view the substance while radio-active is continually being transformed from one state to another. These transformations may be accompanied by the liberation of sufficient energy to supply that carried off by the rays it emits while radio-active. The very large amount of energy emitted by radio-active substances is strikingly shown by some recent experiments of the Curies on the salts of radium. They find that those salts give out so much energy that the absorption of this by the salt itself is sufficent to keep the temperature of the salt permanently above that of the air by a very appreciable amount—in one of their experiments as much as 1.5° C. It appears from their measurements that a gram of radium gives out enough energy per hour to raise the temperature of its own weight of water from the freezing to the boiling point. This evolution of energy goes on uninterruptedly and apparently without diminution. If, however, the views we have just explained are true, this energy arises from the transformation of radium into other forms of matter, and its evolution must cease when the stock of radium is exhausted; unless, indeed, this stock is continually being replenished by the transformation of other chemical elements into radium.

We may make a rough guess as to the probable duration of a sample of radium by combining the result that a gram of radium gives out 100 calories per hour with Rutherford's result that the α rays are particles having masses comparable with the mass of an atom of hydrogen projected with a velocity of about 2×10^9 centimetres per second; for let us suppose that the heat measured by the Curies is due to the bombardment of the radium salt by these particles, and to get a superior limit to the time the radium will last, let us make the assumption that the whole of the mass of radium gets transformed into the α particles (as a matter of fact we know that the emanation is produced as well as the α particles). Let x be the life in hours of a gram of radium; then since the gram emits per hour 100 calories, or 4.2×10^9 ergs, the amount of energy emitted by the radium during its life is $x \times 4.2 \times 10^9$ ergs. If N is the number of α particles emitted in this time, m the mass of one of them in grams, v the velocity, then the energy in the α particles is $\frac{1}{2} N m v^2$, but this is to be equal to $x \times 4.2 \times 10^9$ ergs, hence $\frac{1}{2} N m v^2 = x \times 4.2 \times 10^9$; but if the gram of radium is converted into the α particles, $Nm = 1$, and by Rutherford's experiments $v = 2 \times 10^9$, hence we have $x = \frac{1}{2} \frac{4 \times 10^{18}}{4.2 \times 10^9} = \frac{10^9}{2.1}$ hours, or about 50,000 years.

From this estimate we should expect the life of a piece of radium to be of the order of 50,000 years. This result shows that we could not expect to detect any measurable changes in the space of a few months. In the course of its life the gram of radium will have given out about 5×10^{10} calories, a result which shows that if this energy is derived from transformations in the state of the radium, the energy developed in these transformations must be on a very much greater scale than that developed in any known chemical reactions. On the view we have taken the difference between the case of radium and that of ordinary chemical reactions is that in the latter the changes are molecular, while in the case of radium the changes are atomic, being of the nature of a decomposition of the elements. The example given on page (111) shows how large an amount of energy may be stored up in the atom if we regard it as built up of a number of corpuscles.

We may, I think, get some light on the processes going on in radium by considering the behavior of a model atom of the kind described on page 124, and which may be typified by the case of the corpuscles which when rotating with a high velocity are stable when arranged in a certain way, which arrangement becomes unstable when the energy sinks below a certain value and is succeeded by another configuration. A top spinning about a vertical axis is another model of the same type. This is stable when in a vertical position if the kinetic energy due to its rotation exceeds a certain value. If this energy were gradually to decrease, then, when it reached the critical value, the top would become unstable and would fall down, and in so doing would give a considerable amount of kinetic energy.

Let us follow, then, the behavior of an atom of this type, *i.e.*, one which is stable in one configuration of steady motion when the kinetic energy of the corpuscles exceeds a certain value, but becomes unstable and passes into a different configuration when the kinetic energy sinks below that value. Suppose now that the atom starts with an amount of kinetic energy well above the critical value, the kinetic energy will decrease in consequence of the radiation from the rapidly moving corpuscles; but as long as the motion remains steady the rate of decrease will be exceedingly slow, and it may be thousands of years before the energy approaches the critical value. When it gets close to this value, the motion will be very easily disturbed and there will probably be considerable departure from the configuration for steady motion accompanied by a great increase in the rate at which kinetic energy is loss by radiation. The atom now emits a much greater number of rays and the kinetic energy rapidly approaches the critical value; when it reaches this value the crash comes, the original configuration is broken up, there is a great decrease in the potential energy of the system accompanied by an equal increase in the kinetic energy of the corpuscles. The increase in the velocity of the corpuscles may cause the disruption of the atom into two or more systems, corresponding to the emission of the α rays and the emanation.

If the emanation is an atom of the same type as the original atom, *i.e.*, one whose configuration for steady motion depends on its kinetic energy, the process is repeated for the emanation, but in a very much shorter time, and is repeated again for the various radio-active substances, such as the induced radio-active substance formed out of the emanation.

We have regarded the energy emitted by radium and other radio-active substances as derived from an internal source, *i.e.*, changes in the constitution of the atom; as changes of this kind have not hitherto been recognized, it is desirable to discuss the question of other possible sources of this energy. One source which at once suggests itself is external to the radium. We might suppose that the radium obtained its energy by absorbing some form of radiation which is passing through all bodies on the surface of the earth, but which is not absorbed to any extent by any but those which are radio-active. This radiation must be of a very penetrating character, for radium

retains its activity when surrounded by t
or when placed in a deep cellar. We ar
with forms of Röntgen rays, and of ra
out by radium itself, which can produce
able effects after passing through several
lead, so that the idea of the existence of v
trating radiation does not seem so improb
would have done a few years ago. It is
ing to remember that very penetrating
was introduced by Le Sage more than a
ago to explain gravitation. Le Sage su
that the universe was thronged with exce
small particles moving with very high vel
He called these ultra-mundane corpuscles
sumed that they were so penetrating that
could pass through masses as large as the s
the planets without suffering more than a
slight absorption. They were, however, abso
to a slight extent and gave up to the bo
through which they passed a small fraction
their momentum. If the direction of the u
mundane corpuscles passing through a body
uniformly distributed, the momentum comm
cated by them to the body would not tend to m
it in one direction rather than another, so t
a body A alone in the universe and exposed

bombardment by Le Sage's corpuscles would remain at rest; if, however, there is a second body *B* in the neighborhood of *A*, *B* will shield off from *A* some of the corpuscles moving in the direction *B A*; thus, *A* will not receive as much momentum in this direction as it did when it was alone in the field, but in the latter case it only received enough momentum in this direction to keep it in equilibrium; hence, when *B* is present, the momentum in the opposite direction will get the upper hand so that *A* will move in the direction, *A B*, *i.e.*, will be attracted to *B*. Maxwell pointed out that this transference of momentum from Le Sage's corpuscles to the body through which they were passing involved the loss of kinetic energy by the corpuscles; and that if the loss of momentum were sufficient to account for gravitation, the kinetic energy lost by the ultra-mundane corpuscles would be sufficient, if converted into heat, to keep the gravitating body white hot. The fact that all bodies are not white hot was urged by Maxwell as an argument against Le Sage's theory. It is not necessary, however, to suppose that the energy of the corpuscles is transformed into heat; we might imagine it transformed into a very penetrating radiation which might escape from the gravitating body. A simple calculation will show that the amount of kinetic energy transformed per second in each gram of the gravitating body must be enormously greater than that given out in the same time by one gram of radium.

We have seen in the first chapter that waves of electric and magnetic force possess momentum in their direction of propagation; we might therefore replace Le Sage's corpuscles by very penetrating Röntgen rays. Those, if absorbed, would give up momentum to the bodies through which they pass, and similar consideration to those given by Le Sage would show that two bodies would attract each other inversely as the square of the distance between them. If the absorption of these waves per unit volume depended only upon, and was proportional to, the density, the attraction between the bodies would be directly proportional to the product of their masses. It ought to be mentioned that on this view any changes in gravitation would be propagated with the velocity of light; whereas, astronomers believe they have established that it travels with a very much greater velocity.

As in the case of Le Sage's corpuscles, the loss

of momentum by the Röntgen rays would be accompanied by a loss of energy; for each unit of momentum lost v units of energy would be lost, v being the velocity of light. If this energy were transformed into that of rays of the same type as the incident rays, a little reflection will show that he absorption of the rays would not produce gravitational attraction. To get such attraction the transformed rays must be of a more penetrating type than the original rays. Again, as in the case of Le Sage's corpuscles, the absorption of energy from these rays, if they are the cause of gravitation, must be enormous—so great that the energy emitted by radium would be but an exceedingly small fraction of the energy being transformed within it. From these considerations I think that the magnitude of the energy radiated from radium is not a valid argument against the energy being derived from radiation. The reason which induces me to think that the source of the energy is in the atom of radium itself and not external to it is that the radio-activity of substances is, in all cases in which we have been able to localize it, a transient property. No substance goes on being radio-active for very long. It may be asked how can this statement be reconciled with the fact that thorium and radium keep up their activity without any appreciable falling off with time. The answer to this is that, as Rutherford and Soddy have shown in the case of thorium, it is only an exceedingly small fraction of the mass which is at any one time radio-active, and that this radio-active portion loses its activity in a few hours, and has to be replaced by a fresh supply from the non-radio-active thorium. Take any of the radio-active substances we have described, the *ThX*, the emanations from thorium or radium, the substance which produces induced radio-activity, all these are active for at the most a few days and then lose this property. This is what we should expect on the view that the source of the radio-activity is a change in the atom; it is not what we should expect if the source were external radiation.

CPSIA information can be obtained at www.ICGtesting.com
Printed in the USA
LVOW121547010313

322331LV00010B/415/P